Structural Optimization

William R. Spillers · Keith M. MacBain

Structural Optimization

Springer

William R. Spillers
Department of Civil and Environmental Engineering
New Jersey Institute of Technology
University Heights
Newark, NJ 07102
USA
spillers@njit.edu

Keith M. MacBain
Geiger Engineers
P.C. NY Office
2 Executive Blvd.
Suffern, NY 10901
Suite 410
USA
k_macbain@yahoo.com

Additional material to this book can be downloaded from http://extras.springer.com

ISBN 978-1-4899-7774-8 ISBN 978-0-387-95865-1 (eBook)

DOI 10.1007/978-0-387-95865-1
Springer Dordrecht Heidelberg London New York

Printed on acid-free paper

Springer is part of Springer Science+Business Media (www.springer.com)

for our wives

Preface

Contemporary structural optimization has it roots in the 1960s with Lucien Schmidt's seminal paper.[1] Prior to that time there were no texts on nonlinear programming and if you wanted to do optimization you were relegated to using linear programming. Once mathematical programming was discovered by designers it was thought that engineering design, as an area of study, was over since all you had to do was formulate your design as a nonlinear programming problem and invoke some canned solver. That turned out not to be the case.

While the 1960s and 1970s were characterized by difficulties in solving even small optimization problems (forgetting for the moment optimality criteria methods), the 1990s were characterized by discussions of mathematical programming methods for solving large systems. The flavor of these discussions can be found in the fact that workers in linear programming were then solving *large* systems (Bixby et al., 1991) with, for example, 12 million variables. In fact, today (2007) the web site of Jacek Gondzio describes solving a nonlinear programming problem with 353 million rows and 1010 million columns.

These capabilities of mathematical programming solvers are part of the driving force behind this text: Surely this technology should offer hope to the structural engineer who must commonly deal with large systems. The other factor driving this work is context: We believe that the use of sequential linear programming together with the use of the incremental equations of structures can serve as a focal point about which a quite general structural optimization solver can be developed.

Then there is the question of the availability of software today. In view of the Solver package included in Microsoft EXCEL, an argument can be made that optimization software is now available to everyone. That package is used in this text along with the IMSL routines available with Digital FORTRAN (now sold as Intel FORTRAN). And clearly, all the work described in this text could have as well been done using Matlab. The reader will find a mix of this software used here. The point is that tools are now available for the solution of optimization problems. This includes some freeware available on the Internet that will be discussed later.

With regard to content, this book includes many computer programs most of which are written in FORTRAN. The excuse for focusing so heavily on computing is that contemporary structural optimization is about computing. Parenthetically, we try in this book to do justice to the history of structural optimization but surely have left things out and apologize to those authors whose work has not been properly referenced here.

---

[1] The history of structural optimization has been developed carefully by Wasiutynski and Brandt (1963).

The attempt in writing this book is to help bring the methods of structural optimization into common usage like those of the finite element method. That is, when the structural engineer sits down to design something, he/she should not only have analysis tools available but should also have optimization tools available. In fact, a good case can be made for including these tools in one package since analysis can always be performed as an optimization problem. The engineer could then, for example, automatically change the structural shape as he/she attempts to satisfy some allowable stress constraints. The analysis/redesign cycle could then simply be replaced by optimization steps.

Historically, there has been a tension between proponents of classical optimization methods who claimed that the users of optimality criteria methods were lacking in theory and the users of heuristic schemes such as optimality criteria methods who at the same time claimed that the classical methods were incapable of solving real (large) structures. In view of the tools now available to the engineer, these arguments can be seen to diminish in importance although the optimality criteria methods still have enormous physical appeal.

We attempt to deal fairly with optimality criteria methods but clearly the focus is on the use of sequential linear programming and the incremental equations of structures that comprise a classical approach. The text begins with an Introduction to simple problems of optimization. Then some available tools are discussed. (This second chapter is more formal than the remainder of the text and the first-time reader might treat it lightly.) Chapter 3 introduces our central topic which is structural optimization approached via the incremental equations of structures and sequential linear programming. Chapter 4 then discusses some problems solved using optimality criteria methods. The remainder of the text offers what we see as an overview of the field of structural optimization. This includes beams and plates, dynamic systems, multicriteria methods, and a brief discussion of some ongoing work. The text closes with an Appendix containing three reprints that are regarded by the authors to be basic to the historical development of structural optimization.

Finally, it has been a pleasure to work Springer Science+Business Media and Steven Elliot our editor. For us, Springer has managed to bring the capabilities and competence of an enormous organization to bear without losing the personal touch of staff contacts. You could not ask for more than that.

Keith M. MacBain
Suffern, NY
William Spillers
Newark, NJ

Contents

Contents of the CD

Readme

The readme.txt file contains some basic information as to how to use the CD. More detailed information is given in the text and in the Tools folder.

Computer Programs

This folder contains source code for all programs discussed in the text. A list of these programs is given below.

List of Programs

Program 1. (Prog01.xls) This Excel worksheet is used in Chapter 1 to solve the two-bar truss problem.

Program 2. (Prog02.xls) This Excel worksheet is used in Chapter 1 to solve the frame design.

Program 3. (Prog03.for) This FORTRAN program is used in Section 2.6 of the text as an example of the sequential linear programming solver.

Program 4. (Prog04.xls) This Excel worksheet uses Solver check of the problem solved by Program 3.

Program 5. (Prog05.for) This FORTRAN program is used in Appendix C as a sequential linear programming solver.

Program 6. (Prog06.xls) This Excel worksheet is the Solver solution used in Appendix C.

Program 7. (Prog07.m) This Word file contains a Matlab script file and a Matlab function. They must be separated before they are used. This program is a Matlab version of sequential linear programming.

Program 8. (Prog08.xls) This Excel worksheet is a study of the use of visual basic for structural optimization.

Program 9. (Prog09.for) This FORTRAN program solves a truss with a displacement constraint from Section 3.4 in the text. It is to be used with the data file Prog09.dat.

Program 10. (Prog10.for) This is Program 9 with stress constraints added. It is to be used with the data file Prog10.dat.

Program 11. (Prog11.for) This FORTRAN program solves the classic 25-bar truss problem. It is to be used with the data file Prog11.dat.

Program 12. (Prog12.for) This FORTRAN program solves a plane frame problem with a displacement constraint as discussed in Section 3.7. It is to be used with the data file Prog12.dat.

Program 13. (Prog13.for) This FORTRAN program solves Keller's problem of the optimal shape for a column.

Program 14. (Prog14.for) This FORTRAN program solves the geometric optimization problem discussed in Section 3.9 in the text. It is to be used with the data file Prog14.dat.

Program 15. (Prog15.for and Prog15.dat) This is another version of Program 14.

Program 16. (Prog16.m) This Matlab program solves the problem of an optimal fixed beam as discussed in Section 3.10.

Program 17. (Prog17.m) This Matlab program solves the problem of an optimal fixed plate as discussed in Section 3.11.

Program 18. (Prog18.for) This FORTRAN program solves the most simple iterative design problem for trusses discussed in Section 4.4 of the text. It uses Prog18.dat as a data file.

Program19. (Prog19.for) This FORTRAN program solves the iterative design problem for plane frames discussed in Section 4.7 of the text. It uses Prog19.dat as a data set.

Program 20. Program 20 comprises three programs: Prog20.for, a basic finite element program; Prog21.for, a mesh generator; and Prog22.for, an iterative design program for plane stress problems. The use of these programs is discussed in Section 4.10 in the text.

Program 23. (Prog23.for) This FORTRAN program solves the iterative design problem for a fixed plate as discussed in Section 4.12 of the text.

Program 24. (Prog24.m and Prog25.m) These Matlab programs comprise a program for the solution of the transient response of a tapered beam as discussed in Section 7.2 of the text.

Program 26. (Prog26.for) This FORTRAN program solves the problem of the optimal steady-state vibrations of a truss as discussed in Section 7.3 of the text. It is run with Prog26.dat as a data file.

Program 27. (Prog27.for) This is a incremental solution of the problem solved in Program 26. It uses the same data file.

Program 28. (Prog28.for) This FORTRAN program resolves an example of Farkas from Section 9.3 using AISC stresses. It is to be run with the data file Prog28.dat.

Program 29. (Prog29.for) This FORTRAN program solves the three-dimensional truss. It is used in Problem 3 of Chapter 1. It is to be run with the data file Prog29.dat.

Program 30. (Prog30.m) This Matlab program executes the simplex method. It is an educational program not intended for serious applications.

Program 31. (Prog31.for) This FORTRAN program executes the primal interior point linear programming algorithm described by Arbel. It is to be run with Prog31.dat as a data file.

Program 32. (Prog32.for) This FORTRAN program executes the dual interior point linear programming algorithm described by Arbel. It is to be run with Prog32.dat as a data file.

Program 33. (Prog33.xls) This Excel worksheet is used in Section 9.3 to compute the properties of hollow sections.

Program 34. (Prog34.xls) This Excel worksheet is used in Section 9.3 to compute the properties of angles.

Tools

This folder contains some tools that are considered useful. A list of these tools is given below.

List of Tools

tr2d.	This folder contains an example of using VB code in Excel to automate tasks and using the solver.
g77.zip	A freely available FORTRAN compiler.
MinGW	A freely available FORTRAN compiler. This is the compiler that was tested with the files in optools.zip.
optools	Files that contain IMSL routines used in the text for users without this library.
unixUtils	Unix-like utilities for a PC.
unzip	Executable file to unzip files on the disk.
View3d	Java-based graphical user interface (GUI) framework.

[illegible]

Program 3.2 [illegible] FORTRAN [illegible]
[illegible] with Prog [illegible]
Program 3.3 [illegible]
[illegible]
Program 3.4 [illegible]
[illegible]

1 Introduction

This chapter introduces the reader to various problems that are typical of the field of optimization. It begins with a discussion of structural optimization described formally as a *mathematical programming problem.* This is followed by a series of *typical* optimization problems and an introduction to some of the tools available to deal with them including Lagrange multipliers and linear programming.

1.1 Problem Statement

Structural optimization problems can be deceptively simple to formulate. They can be written as

Find $\boldsymbol{x}$ to minimize $f(\boldsymbol{x})$ subject to $\boldsymbol{g}(\boldsymbol{x}) \leq \boldsymbol{0}$ (1)

Here f (the *objective function)* is a scalar, $\boldsymbol{x}$ is an n-vector (has n components), and $\boldsymbol{g}$ (the *constraints)* is an m-vector. Problems of this type are called *mathematical programming problems* (Luenberger 1984). Equation (1) is typically simplified to read

$$\min f(\boldsymbol{x}) \quad \text{subject to} \quad \boldsymbol{g}(\boldsymbol{x}) \leq \boldsymbol{0} \qquad \text{or even} \qquad \min f(\boldsymbol{x}) \mid \boldsymbol{g}(\boldsymbol{x}) \leq \boldsymbol{0}$$

Note that

$$\boldsymbol{g}(\boldsymbol{x}) \leq 0 \quad \Rightarrow \quad \begin{cases} g_1 \leq 0 \\ g_2 \leq 0 \\ \vdots \\ g_m \leq 0 \end{cases}$$

The specific form of this optimization problem statement is not important since minimize $f \Leftrightarrow$ maximize $-f$ and $\boldsymbol{g}(\boldsymbol{x}) \leq \boldsymbol{0} \Leftrightarrow -\boldsymbol{g}(\boldsymbol{x}) \geq \boldsymbol{0}$ (see Fig. 1.1).

W.R. Spillers, K.M. MacBain, *Structural Optimization*,
DOI 10.1007/978-0-387-95865-1_1, © Springer Science+Business Media, LLC 2009

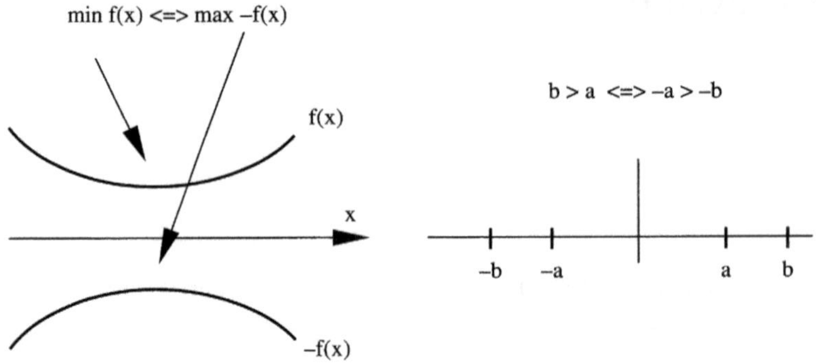

Fig. 1.1. Equivalents in Optimization

1.2 An Optimization Problem

Optimization terminology can be difficult for those unfamiliar with it. This chapter presents some *common* optimization problems and approaches in an attempt to smooth the entrance of the engineer into this world of optimization. The following example will introduce some of the features of mathematical programming problems (Fox 1969).

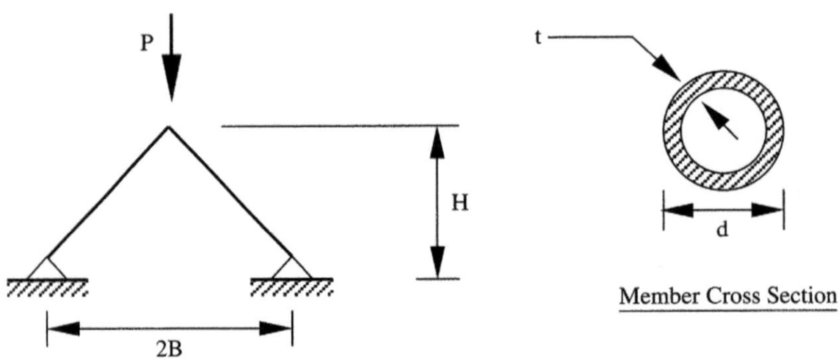

Fig. 1.2. Two-Bar Truss

In this case (Fig. 1.2) the height H and the diameter d of the (tubular) members of a plane two-bar truss are varied in order to minimize the volume of material (which is proportional to the total weight). Additionally, there are two requirements (constraints) that are to be satisfied: (1) the member stress should be less than the yield stress, F_y, and (2) the members should not buckle. Parameters for this problem are tabulated below:

Parameter	Description	Value
E	Young's modulus	29,000 ksi
B	Half-distance between supports	100 in.
F_y	Yield stress of material	36 ksi
t	Wall thickness of tube	0.25 in.
P	Applied load	100 k

Given this problem statement, the objective function and the constraints can be expressed in a mathematical form. Using elementary analysis, the dependent parameters are tabulated below:

Item	Equation
Second moment of inertia (in.⁴)	$I = \frac{\pi}{64}\left[(d+t)^4 - (d-t)^4\right] = \frac{\pi t d}{8}(d^2 + t^2)$
Member force (k)	$F = \frac{P}{2}\frac{\sqrt{B^2 + H^2}}{H}$
Member stress (ksi)	$\sigma = \frac{F}{A}$
Buckling stress (ksi)	$\sigma_{cr} = \frac{\pi^2 EI}{L^2}\frac{1}{A}$

The volume of material is simply the bar area, A, multiplied by the total length of two bars. Using the approximation $A = \pi t d$ (valid for thin-wall tubes), the volume is simply

$$V = 2AL$$
$$= 2(\pi t d)\sqrt{H^2 + B^2}$$

Formally, the objective function, f, and the constraints, $\mathbf{g}$, can now be written as

$$f = 2(d\,t\,\pi)\sqrt{H^2 + B^2}$$

$$g_1 = \frac{P}{2}\frac{\sqrt{H^2 + B^2}}{H}\frac{1}{d\,t\,\pi} - F_y \le 0$$

$$g_2 = \frac{P}{2}\frac{\sqrt{H^2 + B^2}}{H}\frac{1}{d\,t\,\pi} - \frac{\pi^2\,E\left(d^2 + t^2\right)}{8\left(H^2 + B^2\right)} \le 0$$

It is worth emphasizing at this point that *all* optimization problems in this text will be expressed in a similar manner. That is, there is a scalar function f to be optimized subject to zero or more constraints, $\mathbf{g}$.

Figure 1.3 is instructive. It shows a plot of the two constraint surfaces (curves) and curves of constant weight. The curves of constant weight are sometimes called level curves and the figure may be viewed as a topographic map of the solution space.

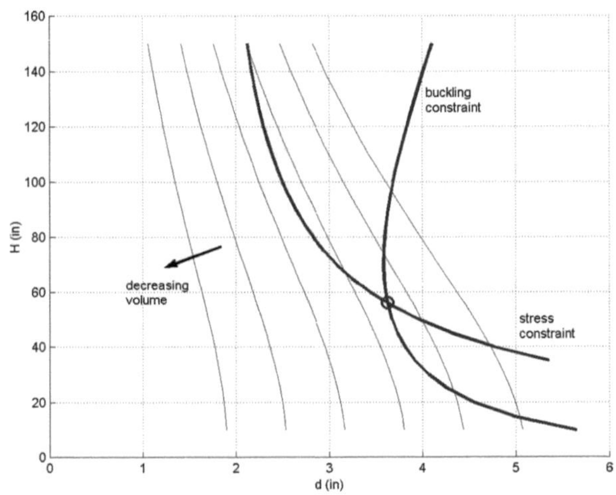

Fig. 1.3. Optimization Solution Space

In this case, the solution is found by starting in the upper right area at a *feasible* point and moving as far "downhill" as possible without passing the "fence" of the constraints.

Using the numerical values given above, this problem can be written as find d and H to

$$minimize \quad f = \frac{\pi}{2} d \sqrt{H^2 + 100^2}$$

subject to

$$50 \frac{\sqrt{H^2 + 100^2}}{H} \frac{4}{d\pi} - 36 \le 0$$

$$50 \frac{\sqrt{H^2 + 100^2}}{H} \frac{4}{d\pi} - 35{,}777 \frac{\left(d^2 + 0.25^2\right)}{\left(H^2 + 100^2\right)} \le 0$$

This problem has a unique solution (not all problems do), which may be found by several methods, many of which will be explored in this text. In this case, the "Solver" from Microsoft Excel was used (see Fig. 1.4.) The solution to this problem is at H = 56 in., d = 3.6 in. More detail on this example and on the use of the Excel Solver can be found in Appendix A and later in this text.

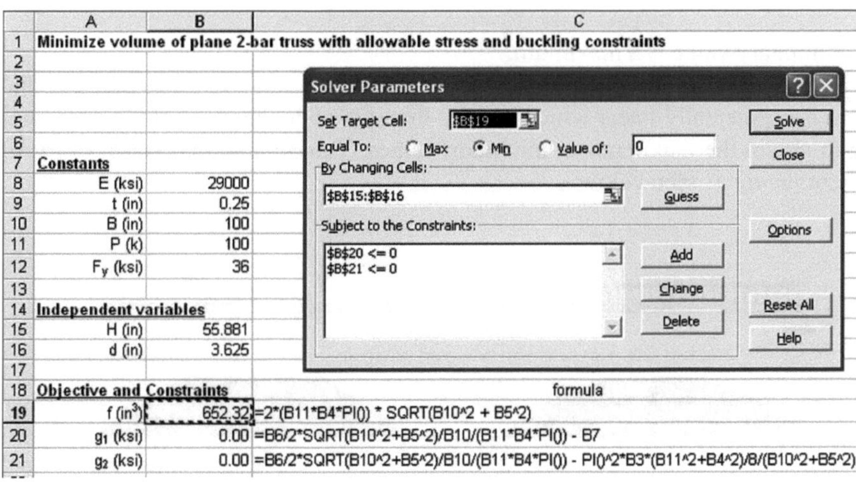

Fig. 1.4. Excel Spreadsheet for Two-Bar Truss Optimization

This spreadsheet is included on the disc as Program 1.

1.3 Elementary Calculus

The engineer typically first encounters optimization in calculus where the derivative is used to identify a stationary point. The second derivative is, of course, required to determine whether a stationary point is a maximum or a minimum. For example (Fig. 1.5), the moment in a uniformly loaded simply supported beam is given by

$$M(x) = \frac{wlx}{2} - \frac{wx^2}{2}$$

Setting the first derivative equal to zero gives a stationary value as

$$\frac{dM}{dx} = M' = \frac{wL}{2} - wx = 0 \quad \Rightarrow \quad x = \frac{L}{2}$$

Since the second derivative is negative,

$$\frac{d^2M}{dx^2} = M'' = -w < 0$$

the stationary value is a maximum.

Figure 1.5 also shows another case of a loaded beam. Here the moment diagram is segmentally linear which implies that it is no longer possible to differentiate to obtain the maximum or minimum values. This situation is typical of *linear programming* as seen below.

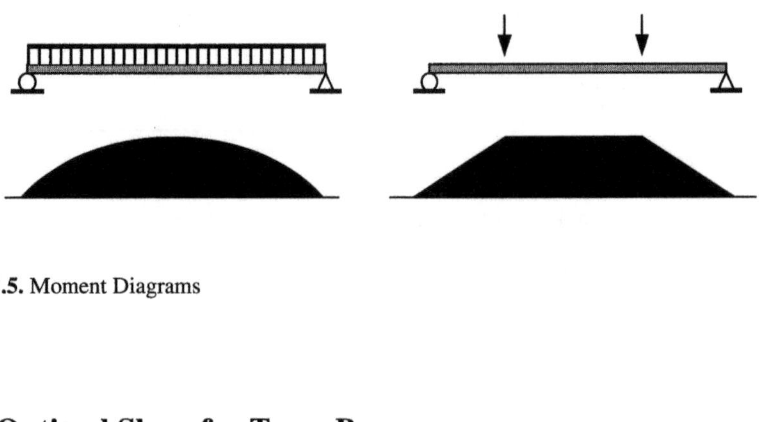

Fig. 1.5. Moment Diagrams

1.4 Optimal Slope for Truss Bars

Given a two-bar truss (Fig. 1.2), it is possible to ask for the optimal bar slope $\theta = \tan^{-1} H/B$, this time using only a single stress constraint. Using the same notation given earlier with H as the (only) independent variable, the elementary calculus approach just presented can be used to solve for H that minimizes the volume. The total volume of the structure is expressed as

$$
\begin{aligned}
V &= 2AL \\
&= 2\left(\frac{F}{\sigma_a}\right)\sqrt{H^2+B^2} \\
&= 2\left(\frac{P}{2}\frac{\sqrt{H^2+B^2}}{H}\frac{1}{\sigma_a}\right)\sqrt{H^2+B^2} \\
&= \frac{P}{\sigma_a}\frac{\left(H^2+B^2\right)}{H}
\end{aligned}
$$

The first derivative with respect to the height H then gives

$$\frac{dV}{dH} = \frac{P}{\sigma_a}\frac{H^2 - B^2}{H^2}$$

and the stationary point is found by setting the first derivative equal to zero.

$$\frac{dV}{dH} = 0 \quad \Rightarrow \quad H = B$$

As before, the sign of the second derivative indicates the type of stationary value, in this case it is positive indicating that the solution is indeed a minimum. Additionally, referring again to Fig. 1.2, $H = B \Rightarrow \theta = 45°$. This example restates what structural engineers know: that truss bars with a flat slope are generally not efficient.

1.5 An Arch Problem

If a parabolic cable under uniform load is inverted, the shape formed is a funicular arch (Fig. 1.6). The following problem asks for the optimal height when the design is done under conditions of a constant allowable stress σ. Using the geometry shown in the figure, $y = 4hx^2/L^2 \Rightarrow y' = 8hx/L^2$.

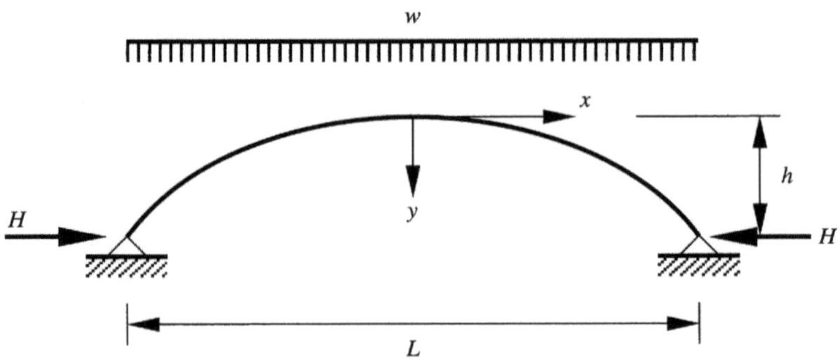

Fig. 1.6. A Parabolic Arch

From symmetry, the vertical reactions are $wL/2$. A free body diagram of ½ the arch can be used to compute the horizontal reaction $H = wL^2/8h$. If a cut is made at any point x, the vertical force component on the cross section can be determined to be wx. The resultant force (axial) at any point x is then $F = [H^2 + (wx)^2]^{1/2}$ and the area $A = F/\sigma$. The volume of material is then

$$V = \int A ds = 2\int_0^{L/2} \frac{1}{\sigma}\left[\left(\frac{wL^2}{8h}\right)^2 + (wx)^2\right]^{1/2}\left[1+\left(\frac{8hx}{L^2}\right)^2\right]^{1/2} dx$$

Since the arc length $ds = [1 + (y')^2]^{1/2}\, dx$, the explanation of the integral is complete. With some rearranging of terms, the square roots cancel and the volume is found to be

$$V = \frac{2w}{\sigma}\left(\frac{L^3}{16h} + \frac{hL}{3}\right)$$

Setting $dV/dh = 0$ gives $h_{opt} = \sqrt{3}\, L/4$ (Haftka, et al. 1990).

1.6 The Gradient of a Function

Given a scalar-valued function of an n-vector $\mathbf{x}$, say $f(\mathbf{x})$, f is said to define a *surface*. The gradient, ∇f, of this scalar function is the vector

$$\nabla f = \begin{bmatrix} \partial f / \partial x_1 \\ \partial f / \partial x_2 \\ \cdot \\ \cdot \\ \partial f / \partial x_n \end{bmatrix}$$

The gradient commonly appears in optimization. It can be used, for example, to describe the variation df of a function as

$df = \nabla f^T dx = \nabla f \cdot dx$ (using a hybrid notation)

with

$$dx = \begin{bmatrix} dx_1 \\ dx_2 \\ \cdot \\ \cdot \\ dx_n \end{bmatrix}$$

This is just the chain rule of partial differentiation since

$$\nabla f \cdot dx \equiv \frac{\partial f}{\partial x_1} dx_1 + \frac{\partial f}{\partial x_2} dx_2 + \cdots + \frac{\partial f}{\partial x_n} dx_n$$

It follows that if $\boldsymbol{n}$ is a unit vector, the directional derivative (in the direction of $\boldsymbol{n}$) is

$$\partial f / \partial n = \nabla f \cdot n$$

Also, ∇f is normal to surfaces of constant f since

$$df = \nabla f \cdot dx = 0 \;\Rightarrow\; \nabla f \perp dx$$

∇f also points uphill in the direction of steepest ascent ($\nabla f \cdot n$ is maximum when ∇f parallels n). For example, let $f = x^2 + y^2 + z^2$. The gradient of f is then

$$\nabla f = \begin{bmatrix} 2x \\ 2y \\ 2z \end{bmatrix}$$

Contours of f are given in Fig. 1.7 along with the gradient at a point and an arbitrary constraint.

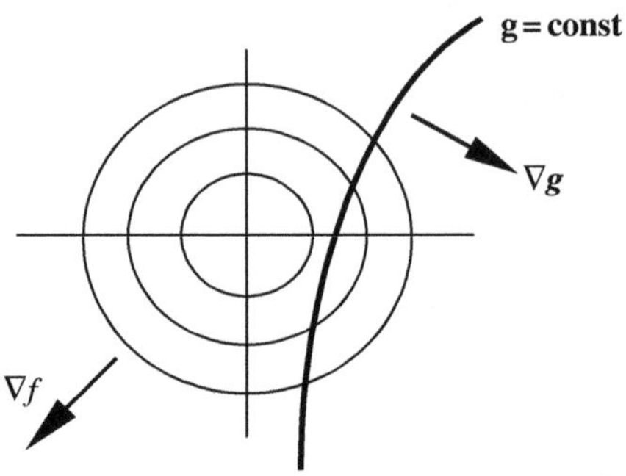

Fig. 1.7. Gradient of a Function

An unconstrained optimization problem can be written using the gradient as $\nabla f = 0$. For the case of f described above, this gives the minimum of f at the point $x = y = z = 0$.

Later the gradient of a *vector* will also be used. For example, consider a problem with equality constraints, $\mathbf{g(x)=0}$. At some point $\mathbf{x}$ which satisfies these constraints there should be a small region $\mathbf{dx}$ which also satisfies the constraints. It follows that in this region $\mathbf{\nabla g \cdot dx = 0}$. The gradient of a vector is a matrix that must be interpreted so that its components represent the chain rule of partial differentiation in the same manner as the scalar case so that $\mathbf{dg = \nabla g \cdot dx}$ can be interpreted again as an application of the chain rule of partial differentiation

$$\nabla g = \begin{bmatrix} \partial g_1/\partial x_1 & \partial g_1/\partial x_2 & \cdots & \partial g_1/\partial x_n \\ \partial g_2/\partial x_1 & \partial g_2/\partial x_2 & \cdots & \\ \vdots & & & \\ \partial g_m/\partial x_1 & \partial g_m/\partial x_2 & \cdots & \partial g_m/\partial x_m \end{bmatrix}$$

1.7 The Lagrange Multiplier Rule

Lagrange multipliers are a basic tool of optimization and will receive considerable attention in this text. However, for the moment it is convenient to follow the approach of Kaplan (1953) in advanced calculus who argues informally that the gradient of the objective function and the gradient of the constraints must be parallel at an optimal point (see Fig. 1.8). He looks at the problem of

$$\max f = xy \quad \text{subject to} \quad g = x^2 + y^2 - 1 = 0$$

This problem can be solved using the so-called *Lagrange multiplier rule* in which the Lagrangian L is formed as

$$L = f + \lambda g = xy + \lambda (x^2 + y^2 - 1)$$

Here L is simply the sum of the objective function plus the constraint multiplied by the constant λ. (λ is called a Lagrange multiplier.) In general, λ will be a vector with m terms corresponding to each of the m constraints of the problem. The optimal solution is then found by differentiating with respect to the variables x, y, and λ as

$$\frac{\partial L}{\partial x} = y + 2\lambda x$$
$$\frac{\partial L}{\partial y} = x + 2\lambda y$$
$$\frac{\partial L}{\partial \lambda} = x^2 + y^2$$

and setting each of these equations equal to zero. Regardless of the value of λ, it follows that $x^2 = y^2$, which gives the four points shown in Fig. 1.8. More consideration will be given to the details in Section 2, particularly with regard to the sign of λ and the character of the problem (i.e., maximization or minimization); however, at this point we note that $x = y = \pm 1$ are solutions to this problem because they do indeed maximize f.

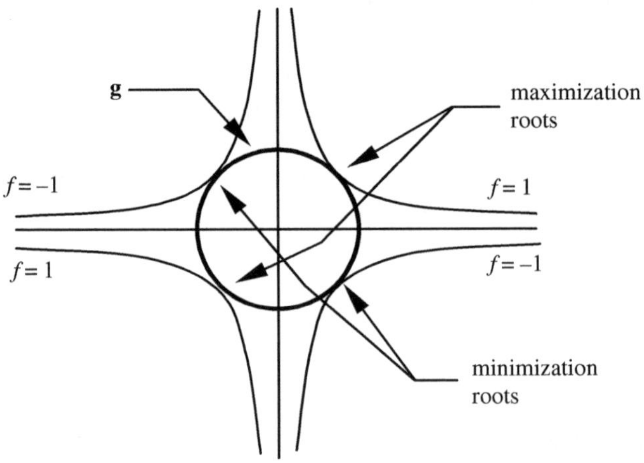

Fig. 1.8. Use of Lagrange Multipliers

In general it will be seen later that the Lagrange multiplier method converts an optimization problem with *equality constraints* into a system of nonlinear equations. It can be seen in the example above that the solution of the nonlinear system can lead to some spurious results so that the analyst must be careful when interpreting his/her results.

1.8 Newton's Method

One of the first topics of numerical analysis is Newton's method. It is generally discussed about the problem of finding roots of an equation such as $f = 0$ but as noted in Sections 1.3 and 1.7, optimization problems are often solved by finding the root to some equation (e.g., $y' = 0$). For example, the buckling load of a column that is fixed at one end and pinned at the other (Timoshenko 1936) requires the solution of the equation $f = x - \tan(x) = 0$ shown in Fig. 1.9.

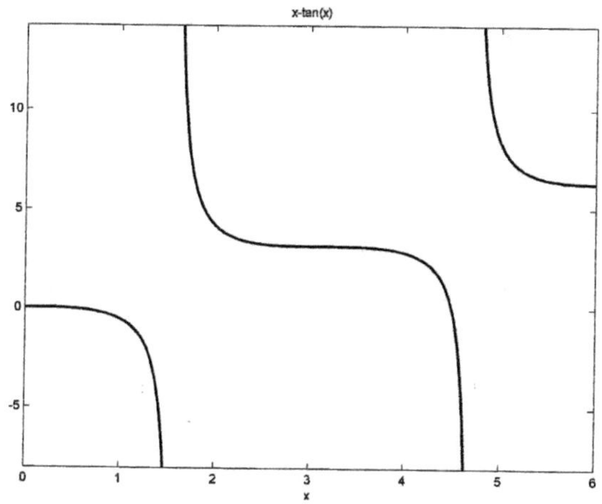

Fig. 1.9. The Function x-tan(x)

That can be done using Newton's method, which is an iterative scheme based on the first two terms of a (linearized) Taylor series to give an estimate of the change in the current estimate, *dx*

$$f \cong f_0 + f'_0\,dx \qquad => \qquad dx = -f_0/f'_0$$

More details on Newton's method is found in Chapter 2; however, here we simply present the iterative equation

$$x_{i+1} = x_i + dx \qquad => \qquad x_{i+1} = x_i - f_i/f'_i$$

where one would typically iterate until the term f_i approaches zero (i.e., the solution is reached). For this example, $f' = 1 - \sec^2(x)$. Starting at $x = 4.4$, the solution proceeds as indicated below:

x	x–tan(x)	1–sec^2(x)	dx
4.4	1.3037	–9.5872	0.1360
4.5360	–1.0738	–31.4692	–0.0341
4.5019	–0.1777	–21.8982	–0.0081
4.4937	–0.0068	–20.2548	–0.0003
4.4934			

Notice that in this case the starting solution is rather close to Timoshenko's solution of $x = 4.493$. That is due to the fact that this problem is rather singular (i.e., asymptotic). In other words, the solution is dependent on the initial solution and an initial value of, say, x = 8, would give a different solution for this problem (verify solution to be x = 10.904). Newton's method will be used heavily in this text, particularly in the so-called method of sequential linear programming described in Chapter 2. This example is included in the Excel file "Chapter 1.xls".

1.9 Solving Linear Equations

Given a system of linear equations, $\boldsymbol{A}\,\boldsymbol{x} = \boldsymbol{b}$, there are three possibilities: a unique solution, many solutions, or no solution. For the case of no solution it is possible to select the vector $\boldsymbol{x}$ so that the norm of the error, **e**, is *minimized.* Let $\boldsymbol{e} \equiv \boldsymbol{b} - \boldsymbol{Ax}$ and the norm of $\boldsymbol{e}$ be defined to be $|\boldsymbol{e}| = (\boldsymbol{e}^{\mathrm{T}}\boldsymbol{e})^{1/2}$. Then

$$\begin{aligned} e^T e &= (b - Ax)^T (b - Ax) \\ &= b^T b - x^T A^T b - b^T Ax + x^T A^T Ax \end{aligned}$$

Differentiating with respect to $\boldsymbol{x}$ to find a stationary point gives

$$\frac{\partial(e^T e)}{\partial x} = 0 \quad \Rightarrow \quad A^T Ax = A^T b \quad \Rightarrow \quad x = (A^T A)^{-1} A^T b$$

The matrix $(\boldsymbol{A}^{\mathrm{T}}\boldsymbol{A})^{-1}\boldsymbol{A}^{\mathrm{T}}$ is sometimes called the *generalized inverse* of the matrix $\boldsymbol{A}$.

There are, of course, many different norms that can be used to measure error. For example, it is common in structures to minimize error using an *energy* or matrix norm with $|\boldsymbol{e}| \equiv (\boldsymbol{e}^{\mathrm{T}}\boldsymbol{Ke})^{1/2}$ and $\boldsymbol{K}$, a positive-definite matrix. It follows that in this case

$$A^T KAx = A^T Kb \quad \Rightarrow \quad x = (A^T KA)^{-1} A^T Kb$$

1.10 Linear Systems Versus Optimization

It is frequently possible to replace a linear system by an optimization problem. For example, given a set of linear equations $\boldsymbol{Ax} = \boldsymbol{b}$, with the matrix $\boldsymbol{A}$ positive definite, an equivalent optimization problem is

minimize $\Phi(\boldsymbol{x}) = ½\,\boldsymbol{x}^{\mathrm{T}}\boldsymbol{Ax} - \boldsymbol{x}^{\mathrm{T}}\boldsymbol{b}$

To see this, consider the exact solution $\boldsymbol{x}^*$.

Let $\boldsymbol{x}^* \mid \boldsymbol{A}\,\boldsymbol{x}^* = \boldsymbol{b}$ then $\boldsymbol{x}^{*\mathrm{T}}\boldsymbol{A}\,\boldsymbol{x}^* = \boldsymbol{x}^{*\mathrm{T}}\boldsymbol{b}$

Now consider $\boldsymbol{x} = \boldsymbol{x}^* + \boldsymbol{\varepsilon}$, which differs from the exact solution, $\boldsymbol{x}^*$, by some small amount $\boldsymbol{\varepsilon}$. The objective function Φ evaluated at $\boldsymbol{x}$ is then

$$\begin{aligned}\Phi(x) &= \tfrac{1}{2}\,(\mathbf{x}^*+\boldsymbol{\varepsilon})^T\mathbf{A}(\mathbf{x}^*+\boldsymbol{\varepsilon}) - (\mathbf{x}^*+\boldsymbol{\varepsilon})^T\mathbf{b} \\ &= \tfrac{1}{2}\,(\mathbf{x}^*)^T\mathbf{A}(\mathbf{x}^*) + \tfrac{1}{2}\,(\boldsymbol{\varepsilon})^T\mathbf{A}(\boldsymbol{\varepsilon}) + (\boldsymbol{\varepsilon})^T\mathbf{A}(\mathbf{x}^*) - (\mathbf{x}^*)^T\mathbf{b} - (\boldsymbol{\varepsilon})^T\mathbf{b} \\ &= \Phi(\mathbf{x}^*) + \text{positive quantity}\end{aligned}$$

Since $\boldsymbol{\varepsilon}^T(\mathbf{A}\mathbf{x}^* - \mathbf{b}) = 0$ It follows that

$$\text{minimize } \Phi(\boldsymbol{x}) = \tfrac{1}{2}\,\boldsymbol{x}^T\boldsymbol{A}\boldsymbol{x} - \boldsymbol{x}^T\boldsymbol{b} \qquad \Leftrightarrow \qquad \mathbf{A}\mathbf{x} = \mathbf{b}$$

1.11 Equations of Structures

It is convenient here to introduce the equations of discrete structures as they are to be used in this book. They can be contrasted with the equations of elasticity as indicated.

	Structures	Elasticity
Equilibrium equation:	$\mathbf{N}^T\mathbf{F} = \mathbf{P}$	$\sigma_{ij,i} + f_j = 0$
Constitutive equation:	$\mathbf{F} = \mathbf{K}\,\boldsymbol{\Delta}$	$\sigma_{ij} = 2\mu\varepsilon_{ij} + \lambda\delta_{ij}\varepsilon_{kk}$
Member/node displacement eq. (strain/displacement)	$\boldsymbol{\Delta} = \mathbf{N}\,\boldsymbol{\delta}$	$\varepsilon_{ij} = \tfrac{1}{2}\,(u_{i,j} + u_{j,i})$

Here

- $\mathbf{F}, \boldsymbol{\Delta}$ $(\sigma_{ij}, \varepsilon_{ij})$ – member force, displacement matrix (stress, strain tensor)
- $\mathbf{P}, \boldsymbol{\delta}$ (f, u) – node force, displacement matrix (traction, displ. vector)
- $\mathbf{K}$, (μ, λ) – primitive stiffness matrix (Lame constants)
- $\mathbf{N}$ – generalized incidence matrix

As an example of a structural optimization problem, Fig. 1.10 shows a three-bar truss. In this case the problem is to find the member areas that minimize the weight (volume) while satisfying allowable stress constraints. Here the structural matrices are

$$N = \begin{bmatrix} 1/\sqrt{2} & 1/\sqrt{2} \\ 0 & 1 \\ -1/\sqrt{2} & 1/\sqrt{2} \end{bmatrix} \qquad K = \begin{bmatrix} K_1 & 0 & 0 \\ 0 & K_2 & 0 \\ 0 & 0 & K_3 \end{bmatrix} \qquad P = \begin{bmatrix} P_X \\ P_Y \end{bmatrix}$$

The elastic stiffness matrix, $K_E = N^T K N$, is then

$$K_E = \begin{bmatrix} (K_1 + K_3)/2 & (K_1 - K_3)/2 \\ (K_1 - K_3)/2 & (K_1 + K_3)/2 + K_2 \end{bmatrix}$$

and the inverse stiffness matrix is

$$K_E^{-1} = \begin{bmatrix} (K_1 + K_3)/2 + K_2 & -(K_1 - K_3)/2 \\ -(K_1 - K_3)/2 & (K_1 + K_3)/2 \end{bmatrix} \frac{2}{2K_1K_3 + K_2(K_1 + K_3)}$$

Here $K_I = A_I E/L_I$, with

A_I – area of bar I
E – Young's modulus
L_I – the length of bar I

The member force matrix F can now be written in terms of the member stiffnesses as

$$F = \begin{bmatrix} F_1 \\ F_2 \\ F_3 \end{bmatrix} = \begin{bmatrix} \frac{K_1}{\sqrt{2}}[2(K_1 + K_3)P_X + 2K_3P_Y] \\ K_2[-(K_1 - K_3)P_X + (K_1 + K_3)P_Y] \\ \frac{K_3}{\sqrt{2}}[-2(K_1 + K_2)P_X + 2K_1P_Y] \end{bmatrix} \frac{1}{2K_1K_3 + K_2(K_1 + K_3)}$$

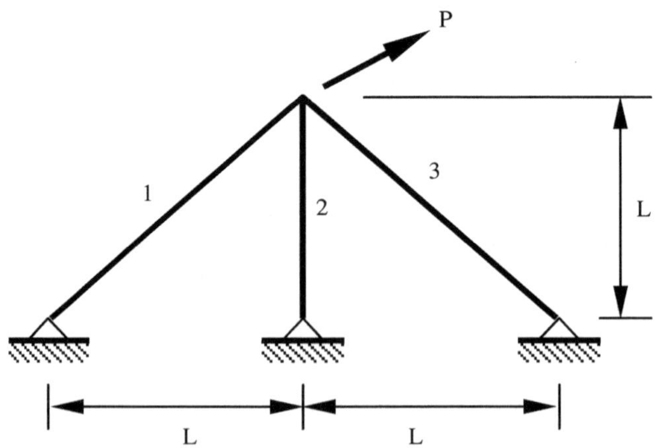

Fig. 1.10. Three-bar Truss

where F_i is the force in bar i. What emerges is the fact that while the stiffness matrix K_E is linear in the member stiffnesses (and therefore in the member areas), the displacements and the member forces are controlled by the inverse of K_E which is a rational function of the member stiffnesses and nontrivial to deal with.

A weight-minimization problem for this truss might be written as follows: Given load matrix P and the structural geometry, find the member areas A_i that minimize the weight while satisfying allowable stress requirements. Formally, find A_1, A_2, A_3 which

$$\text{Minimize } f = A_1L_1 + A_2L_2 + A_3L_3$$
$$\text{Subject to} \quad g_1 = |F_1| - \sigma_1(A_1)\,A_1 \le 0$$
$$g_2 = |F_2| - \sigma_2(A_2)\,A_2 \le 0$$
$$g_3 = |F_3| - \sigma_3(A_3)\,A_3 \le 0$$

Here the σ_i are the allowable stresses which typically vary with the member size and length and the design code used. In the space of areas, surfaces of constant weight (volume) are planes (see Fig. 1.11).

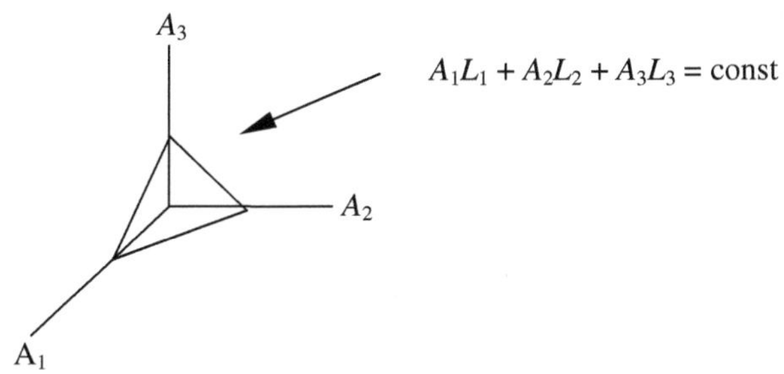

Fig. 1.11. Area Space

The truss problem just described turns out to be nontrivial to solve as a mathematical programming problem. On the other hand, when *optimality criteria methods* are discussed later, a version of this problem will be solved using a very simple algorithm. There are two points to be made here. First, there is an interesting tension between classical optimization and optimality criteria methods, which will be considered later. Second, structural optimization is computationally intensive and it is only recently that the tools have become generally available to solve practical structural optimization problems routinely. Appendix B gives a more careful discussion of the truss problem.

1.12 Plastic Analysis

The simplest approach to plastic analysis (Heyman, 2008 and Spillers, 1985) lies through *linear programming* (optimization with a linear objective function and linear constraints). Given a frame with only proportionally applied joint loads, the lower bound theorem of plastic analysis states that the collapse load is the largest load which is safe and satisfies the requirements of equilibrium. Formally this can be written as

maximize λ such that $N^T F = \lambda P$ (equilibrium)

$|m^+_i| \le (M_{ult})_i$ (safety)

$|m^-_i| \le (M_{ult})_i$ (safety)

for each member i

Where F, P, and N are the structural matrices and λ is a load factor. m_i is the moment at the end of the member, and M_{ult} is the moment capacity of the member. For example, Fig. 1.12 shows a plane frame whose elements have moment capacities of μ (columns) and 2 μ (beams).

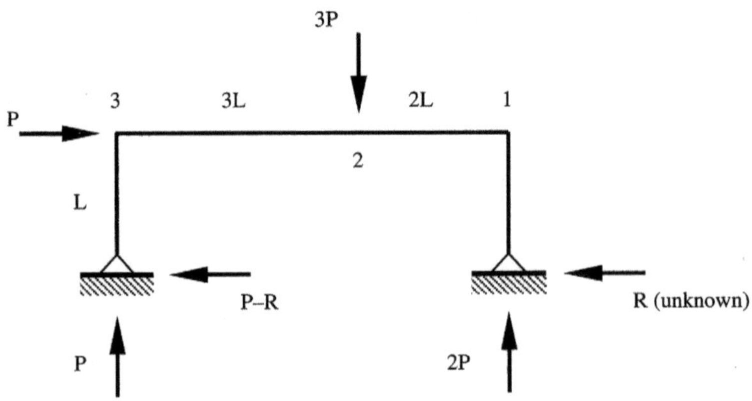

Fig. 1.12 Rigid Frame

There are three possible plastic hinges that may form. These are tabulated below in descriptive and equation form.

Location of hinge	Equation
Top of the right-hand column (node 1)	$RL = \mu \Rightarrow |R| \le \mu/L$

Top of the left-hand column (node 3)	$(P-R)L = \mu \Rightarrow \mid P-R \mid \leq \mu/L$
Under vertical load (node 2)	$2P \times 2L - RL = 2\mu \Rightarrow \mid 4P-R \mid \leq 2\mu/L$

These inequalities are shown in Fig. 1.13.

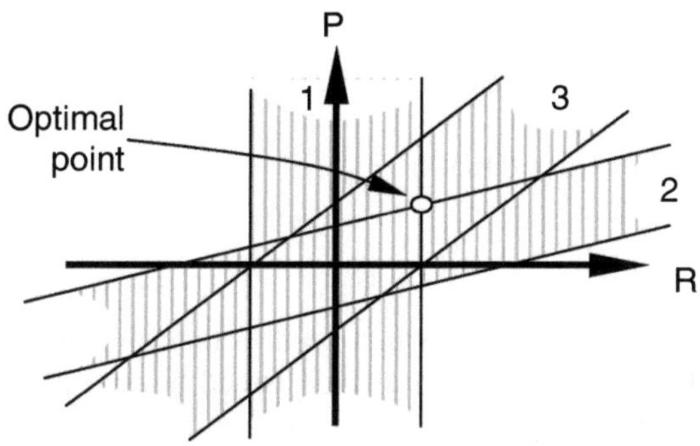

Fig. 1.13 Linear Programming Solution

One may note that in the figure above, constraint 2 limits the upward movement (i.e., increasing P) while constraint 1 limits the movement to the right (i.e., increasing R). In this case the constraint on node 3 does not govern the design. Procedures for numerically solving linear programming problems such as this are given in Chapter 2.

1.13 A Beam Problem

Given a beam with span L and load w, there is a simple optimization problem which asks for the optimal shape of the beam under the assumption that the weight of a beam element at any point is proportional to the absolute value of the bending moment m at that point (after William Prager). In this model, only equilibrium is satisfied. (It can be thought of as a plastic design problem.) The question is then to find m that will

$$\text{minimize} \int |m|\, dx \qquad \text{subject to} \quad m'' = -w \ \text{(equilibrium)}$$

Multiplying the equilibrium equation by a displacement-like quantity and integrating by parts twice give the virtual work equation. (More care will be taken with the end points of the integrals later.)

$$\int wy\, dx = -\int my''\, dx$$

Assume that $|y''| \leq$ some constant k which is taken here to be 1 as a matter of convenience. The following continued inequality then holds:

$$\int |m|\, dx \geq \int |m|\ |y''|\, dx \geq -\int m\, y''\, dx = \int wy\, dx$$

Prager uses this inequality to argue that if

$$y'' = -\operatorname{sgn}(m)$$

equality holds in the above inequality which implies that the subsequent moment diagram is optimal. For example, given a fixed ended beam with a concentrated load at its center, $y'' = \pm 1$ implies that the inflection points lie at the quarter points as shown in Fig. 1.14 and that the minimum value of $\int |m|\, dx$ is $PL^2/16$.

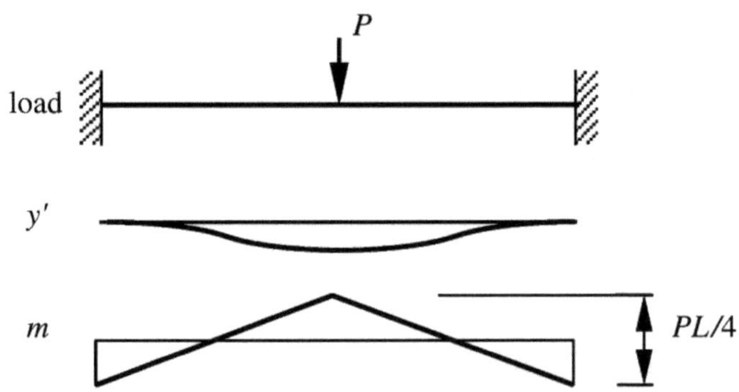

Fig. 1.14 Plastic Design

1.14 Quadratic Programming

One of the most simple results of optimization theory is that problems of quadratic programming (Byrd et al., 2005) with linear equality constraints reduce to the solution of a linear system. That will be demonstrated now for a structures problem. It will be shown that the equations of the node method are equivalent to the system

$$\text{minimize } \tfrac{1}{2} F^T K^{-1} F \qquad \text{subject to} \qquad N^T F = P$$

In this case a structure is given and member forces are sought to satisfy the optimization problem stated. Using the Lagrange multiplier rule, the Lagrangian *L* is formed as

$$L = ½ F^T K^{-1} F + \lambda^T (P - N^T F)$$

The gradient of *L* then gives

$$\partial L/\partial F_I = 0 \qquad \Leftrightarrow \qquad K^{-1}F = N\lambda$$

(Note that $F = K\Delta$ and $\Delta = N\delta$ can be combined to give $K^{-1}F = N\delta$.)

This example is particularly interesting because the Lagrange multiplier (matrix) plays the role of the displacement matrix in the node method. (See the problems at the end of the chapter for another example.)

1.15 Embedding

The late mathematician Richard Bellman coined the term *embedding*. The clever use of embedding involves simplifying a problem by neglecting certain conditions, which, if you are lucky, in the end do not matter. That occurs in this text with structures problems in which compatibility is neglected. In the cases in which the optimal solution is, for example, statically determinate, compatibility becomes a moot issue. (In a statically determinate structure $N^T F = P \Rightarrow F = (N^T)^{-1}P$ and $\delta = N^{-1}\Delta = N^{-1}K^{-1} F$ which implies, for example, that it is possible to construct or *realize* a statically determinate design.) It will also be seen in other cases that the optimal design found through embedding may not be *realizable.* But the unrealizable design can in fact be a good starting point for some reanalysis. In any case, embedding can be a useful tool for optimization.

1.16 Geometric Programming

There is a branch of mathematical programming called *geometric programming* (Duffin et al., 1967) which is roughly based upon the geometric inequality. It is quite lovely to see when it works. The flavor of this activity can be seen in the following example taken from Duffin:

$$\begin{aligned}(U_1 - U_2)^2 \geq 0 \quad &\Rightarrow \quad U_1^2 - 2U_1U_2 + U_2^2 \geq 0 \\ &\Rightarrow \quad ½ u_1 + ½ u_2 \geq u_1^{1/2} u_2^{1/2}\end{aligned}$$

with $u_1 = U_1^2$ and $u_2 = U_2^2$. Consider now the function $g(t) = 4t + 1/t$. Let $½ u_1 = 4t$ and $½ u_2 = 1/t$. A lower bound on g(t) then is $u_1^{1/2} u_2^{1/2} = (8t)^{1/2} (2/t)^{1/2} = 4$. (Notice how t cancels out.) Since equality is achieved when $U_1 = U_2$ it follows that 4t = 1/t or t=1/2 at its minimum.

Morris (1972) discusses applications of geometric programming to structural optimization.

1.17 Plastic Design of Plane Frames

There is a simple model of plastic design in which it is assumed that the area of a member is proportional to the absolute value of the maximum moment that occurs in a beam segment. Using the lower bound theorem of plastic analysis, the plastic design problem for a frame can be written as

$$\min \sum \max\left\{| m_i^+ |, | m_i^- |\right\} L_i \quad \textit{subject to} \quad N^T F = P \textit{ and } | m_i^\pm | \leq \mu_i$$

Where m_i is the moment at the end of member *i* and μ_i is the moment capacity of member *i*. As an example of this consider the frame of Fig. 1.15 with the given loading, dimensions, and node numbering.

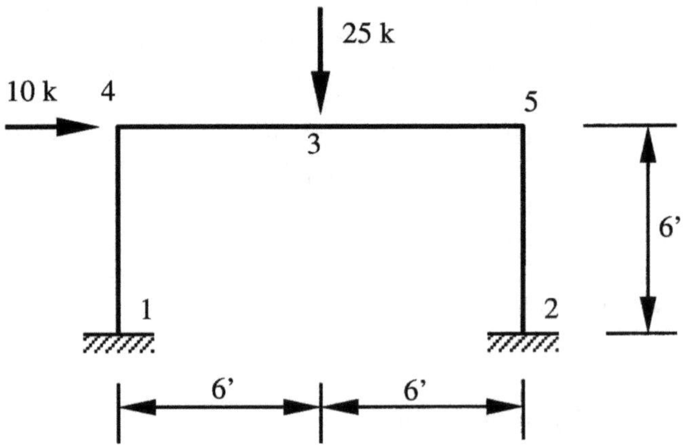

Fig. 1.15. Rigid Frame

This frame is statically indeterminate to the third degree. In order to deal with equilibrium, it is made statically determinate by introducing three hinges and subsequently applying loads at the hinges representing the internal forces that are present in the actual structure. Hinge locations are arbitrarily selected to be at nodes 1, 2, and 3. This gives four load cases on the determinate structure and the solution is simply represented by the superposition of these four. Using a capital letter *M* to represent the moment at a node and the sign convention of positive for compression on the outer face, the remaining internal moments, M_4 and M_5, are expressed for each these four cases in Fig 1.16.

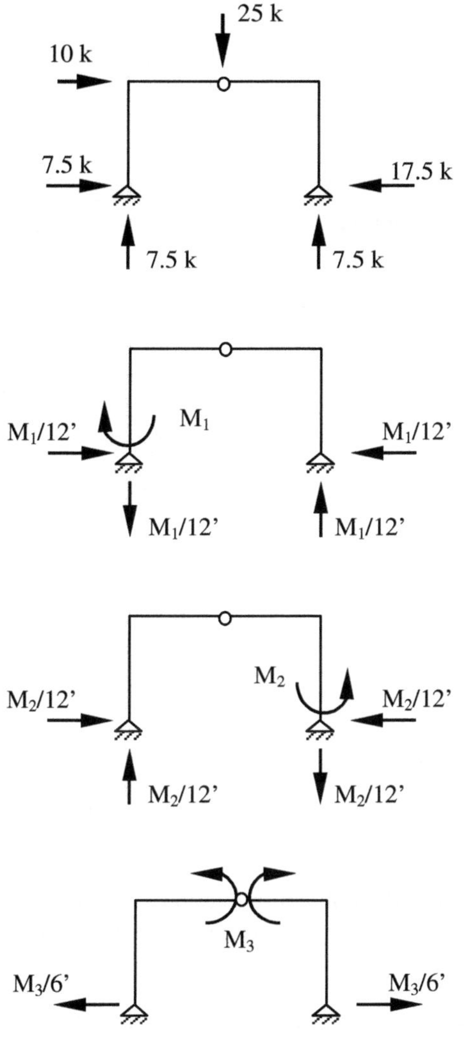

$$M_4 = -45\ k \cdot ft$$
$$M_5 = -105\ k \cdot ft$$

$$M_4 = 0.5\, M_1$$
$$M_5 = -0.5\, M_1$$

$$M_4 = -0.5\, M_2$$
$$M_5 = 0.5\, M_2$$

$$M_4 = M_3$$
$$M_5 = M_3$$

Fig. 1.16. Loadings on Determinate Structure

For this case, the plastic design of frames has the form

$$\min\{\ \max[|M_1|, |M_4|] + \max[|M_3|, |M_4|] + \max[|M_3|, |M_5|] + \max[|M_2|, |M_5|]\ \}$$

subject to

$$M_4 = -45 + M_1/2 - M_2/2 + M_3$$
$$M_5 = -105 - M_1/2 + M_2/2 + M_3$$
$$|M_i| \le \mu_i$$

An Excel Solver solution (Program 2 on the CD) is indicated in Fig. 1.17.

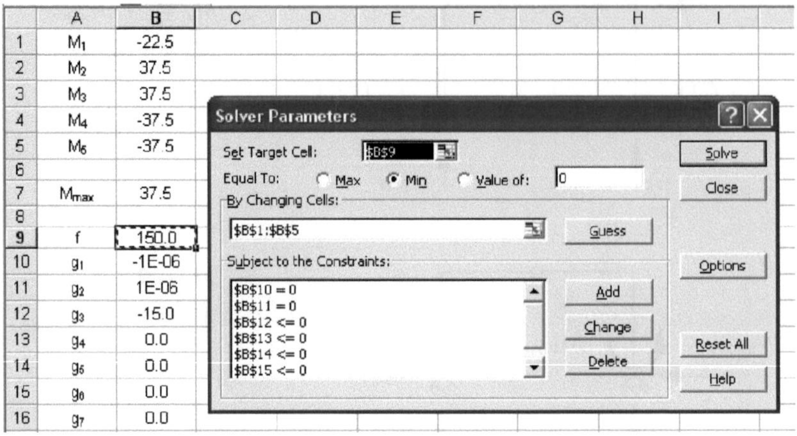

Fig. 1.17. Frame Design Spread Sheet

1.18 Problems

1. Rerun the example in Section 1.2 using Excel for the case in which F_y = 60 ksi.

2. Solve the plastic analysis problem in Section 1.12 for the case in which the beams and columns have the same moment capacity.

3. Annotate fully the program Prog29.for on the disk. Run this program using the file Prog29.dat as data.

4. Find the maximum of the function $f = R^2 - x^2 - y^2$ using the gradient.

5. Solve the problem

   $\min x^2 + y^2 + z^2$ subject to $x + y + z = 1$

$$x{-}y = 1$$

Solution:

In general, min x^TKx | $Ax=b$ => $L= x^TKx + \lambda^T(b-Ax)$

$2Kx=A^T\lambda$ => $x=1/2\ K^{-1}A^T\lambda$ => $b=1/2\ A\ K^{-1}A^T\lambda$

In this case,

$$x = \begin{bmatrix} x \\ y \\ z \end{bmatrix} \qquad A = \begin{bmatrix} 1 & 1 & 1 \\ 1 & -1 & 0 \end{bmatrix} \qquad b = \begin{bmatrix} 1 \\ 1 \end{bmatrix} \qquad K = I$$

$$\lambda = \left(AA^T\right)^{-1} 2b = \begin{bmatrix} 2/3 \\ 1 \end{bmatrix} \qquad x = \frac{1}{2} K^{-1} A^T \lambda = \begin{bmatrix} 5/6 \\ -1/6 \\ 1/8 \end{bmatrix}$$

and the optimal value of the function is 0.833.

6. Consider the minimum weight design of the four-bar truss shown in Fig. 1.18 (after Haftka 1990). The length of members 1–3 is L and the length of member 4 is √3L. For the sake of simplicity we assume that members 1–3 have the same area A_1 and member 4 has area A_2.

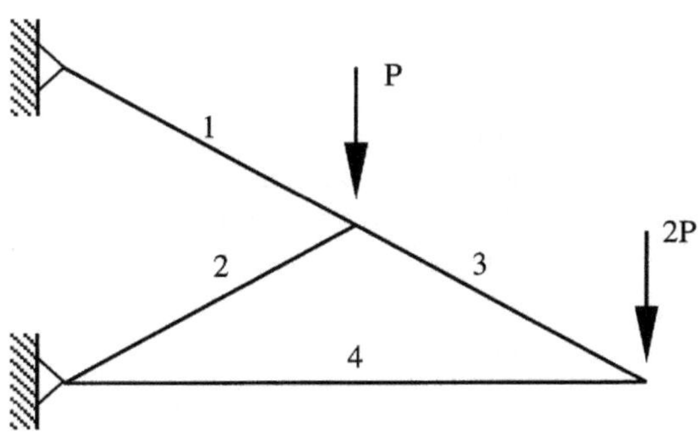

Fig. 1.18. The Truss of Problem 6

The constraints are limits on the stresses in the members and on the vertical displacement at the right end of the truss. Under the specified loading the member forces, F, and the vertical displacement δ at the end can be easily verified to be

$$F_1 = 5P; \quad F_2 = -P; \quad F_3 = 4P; \quad F_4 = -2\sqrt{3}P$$

$$\delta = \frac{6PL}{E}\left(\frac{3}{A_1} + \frac{\sqrt{3}}{A_2}\right)$$

We assume the allowable stresses in tension and compression to be 8.74×10^{-3} E and 4.83×10^{-4} E, respectively, where E is Young's modulus. The vertical displacement is to be no greater than 3×10^{-3} L. We note that the extreme tension stress occurs in member 1; however, because the member areas are not yet determined, it is not known whether the maximum compression stress will occur in member 2 or 4; thus both should be considered at this point. The standard form of this optimization problem follows.

$$\begin{aligned}
&\textit{minimize} && f = 3A_1 + \sqrt{3}\,A_2 \\
&\textit{subject to} && 5\frac{P}{A_1} \le 8.74\times10^{-4}\ E \\
& && \frac{P}{A_1} \le 4.83\times10^{-4}\ E \\
& && 2\sqrt{3}\frac{P}{A_2} \le 4.83\times10^{-4}\ E \\
& && 6\frac{PL}{E}\left(\frac{3}{A_1} + \frac{\sqrt{3}}{A_2}\right) \le 3\times10^{-3}\ L
\end{aligned}$$

From the above, it is clear that the second constraint (i.e., allowable compressive stress on member 2) is redundant and may be removed from the system. Substituting $x_1 = A_1E/(1000\ P)$ and $x_2 = A_2E/(1000\ P)$, this system can be simplified to

$$\begin{aligned}
&\textit{minimize} && f = 3x_1 + \sqrt{3}\,x_2 \\
&\textit{subject to} && 0 \le x_1 - 5.721 \\
& && 0 \le x_1 - 7.172 \\
& && 0 \le 3 - 6\left(\frac{3}{x_1} + \frac{\sqrt{3}}{x_2}\right)
\end{aligned}$$

An Excel solution for this problem is given in Fig. 1.19, where the solution variables x_1, x_2 are shown along with the problem variables A_1, A_2.

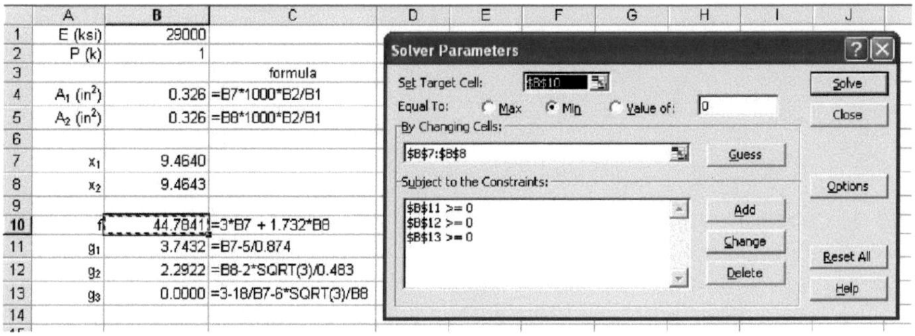

Fig. 1.19 Excel Spreadsheet for Problem 6

7. Use Newton's method to find other roots of the function $f = x - \tan(x) = 0$ given in Section 1.8.

An Excel solution for this problem is [illegible] in Figure [illegible] This solution [illegible] [illegible] variables [illegible]

2 Some Tools of Optimization

This chapter discusses, in more detail, the tools of optimization that were introduced in Chapter 1. It begins with Lagrange multipliers, which see more use in optimization than any other device. It moves on then to discussions of the Kuhn–Tucker conditions and calculus of variations. It includes methods for solving optimization problems numerically such as sequential linear programming, which is the workhorse of this text.

2.1 The Lagrange Multiplier Rule

The Lagrange multiplier rule and the Kuhn–Tucker conditions come directly from topics of linear algebra. Gale (1960) offers a very elegant approach to these topics that is highly recommended by the authors. The approach used here is somewhat more intuitive.

The Lagrange multiplier rule can be used to convert an optimization problem with *equality* constraints into a nonlinear system of equations. It states that under appropriate circumstances there exist Lagrange multipliers, λ, such that

$$\text{minimize } f(x) \mid \mathbf{g}(x) = 0 \quad \Leftrightarrow \quad \nabla f + \lambda^T \nabla g = 0\,,\ g = 0.$$

While the Lagrange multiplier rule itself skirts basic optimization ideas, its proof deals with them directly. The idea is that at an optimal point, the variation of f, $df = \nabla f \cdot dx$, must be zero or positive in any feasible direction dx which must also satisfy $dg = \nabla g \cdot dx = 0$. (Were this not the case it would be possible to reduce the objective function implying that the point in question is not optimal.)

One approach to Lagrange multipliers comes directly from the either/or theorems of linear algebra (Gale 1960),

Either the equation $A^T x = b$ has a solution $x \neq 0$ or the equations $Ay = 0$ and $b^T y = 1$ have a solution.

Obviously, both alternatives cannot hold since $A^T x = b \Leftrightarrow y^T A^T x = y^T b = 0$ if $Ay = 0$ also holds. In terms of optimization this either/or theorem reads

Either the equation $\nabla g^T \lambda = -\nabla f$ has a solution $\lambda \neq 0$ or the equations $\nabla g \cdot dx = 0$ and $\nabla f \cdot dx = -1$ have a solution.

W.R. Spillers, K.M. MacBain, *Structural Optimization*,
DOI 10.1007/978-0-387-95865-1_2, © Springer Science+Business Media, LLC 2009

Mechanically the Lagrange multiplier rule is sometimes described as follows. Given an optimization problem with equality constraints, the *Lagrangian, L,* is formed by combining *f* with each constraint g_i multiplied by a Lagrange multiplier λ_i as

$$L = f + \lambda^T g = f + \sum_{i=1}^{m} \lambda_i g_i$$

This allows a constrained minimization problem to be treated as an unconstrained problem that is then solved by taking the gradient

$$\nabla L = \nabla f + \lambda^T \nabla g = 0 \quad \text{with} \quad g = 0$$

also holding. In more detail this system can be written as

$$\frac{\partial f}{\partial x_j} + \frac{\partial}{\partial x_j} \sum_{i=1}^{m} \lambda_i g_i = 0, \qquad j = 1, \ldots, n$$

Another, rather physical, derivation of the Lagrange multiplier rule follows. Let x be an optimal point so that $df = \nabla f \cdot dx = 0$ for any $\mathbf{dg} = \mathbf{\nabla g} \cdot dx = 0$. Let $\mathbf{\nabla g}$ have rank m. Partition $\mathbf{\nabla g}$**,** perhaps after reordering, so that

$$\nabla g \cdot dx = 0 \qquad \Rightarrow \qquad [\nabla g_1, \nabla g_2] \begin{bmatrix} dx_1 \\ dx_2 \end{bmatrix} = 0$$

with $\mathbf{\nabla} g_i$ nonsingular. It follows that

$$dx_1 = -\nabla g_1^{-1} \nabla g_2 dx_2$$

where dx_2 can be selected arbitrarily. Under these conditions, $\mathbf{\nabla g}^T \lambda = \nabla f$ can be written as

$$\begin{bmatrix} \nabla g_1^T \\ \nabla g_2^T \end{bmatrix} \lambda = \begin{bmatrix} \nabla f_1 \\ \nabla f_2 \end{bmatrix}$$

which can be solved for λ as

$$\lambda = (\nabla g_1^T)^{-1} \nabla f_1$$

It remains to show that

$$\nabla g_2^T \lambda = \nabla f_2 \qquad \Rightarrow \qquad \nabla g_2^T (\nabla g_1^T)^{-1} \nabla f_1 = \nabla f_2$$

Now in partitioned form

$$\nabla f \cdot dx = 0 \qquad \Rightarrow \qquad \nabla f_1 \cdot dx_1 + \nabla f_2 \cdot dx_2 = 0$$

or

$$-\nabla f_1^T \nabla g_1^{-1} \nabla g_2 dx_2 + \nabla f_2 \cdot dx_2 = 0$$

or

$$(\nabla g_2^T (\nabla g_1^T)^{-1} \nabla f_1 - \nabla f_2) \cdot dx_2 = 0$$

Since dx_2 is arbitrary, the term in parenthesis must be zero completing the proof of the theorem. Several examples of the use of Lagrange multipliers are included in Chapter 1.

2.2 The Kuhn–Tucker Conditions

The Kuhn–Tucker conditions, sometimes called the Karush–Kuhn–Tucker conditions, deal with optimization problems with *inequality* constraints. They do for inequality constraints what the Lagrange multiplier theorem does for equality constraints, but inequality constraints turn out to be much more difficult to deal with. The result is that the computational role of the Kuhn–Tucker conditions is nothing as pervasive as is the Lagrange multiplier rule.

Theorem: (Kuhn–Tucker) At an optimal point of the problem minimize f(x) subject to **g(x) ≤ 0**, *there exist Lagrange multipliers* $\lambda \geq 0$ *which satisfy* $\nabla f + \lambda^T \nabla g = 0$ *and* $\lambda_i g_i = 0$ *for i = 1,...,m.*

The proof of this lemma follows directly from

Farkas' Lemma (Simonnard, 1966): The statement that $a \cdot x < 0$ *for all x such that* $Ax \geq 0$ *is equivalent to the statement that there exists a* $\lambda \geq 0$ *such that* $a + A^T\lambda = 0$.

Proof: The proof proceeds by induction on the number of rows of A. When A has one row it is only necessary to show that $a + A^T\lambda = 0 \Leftrightarrow a^T x < 0$, $Ax \geq 0$. This is shown in Fig. 2.1. In this case a, A, and x are vectors.

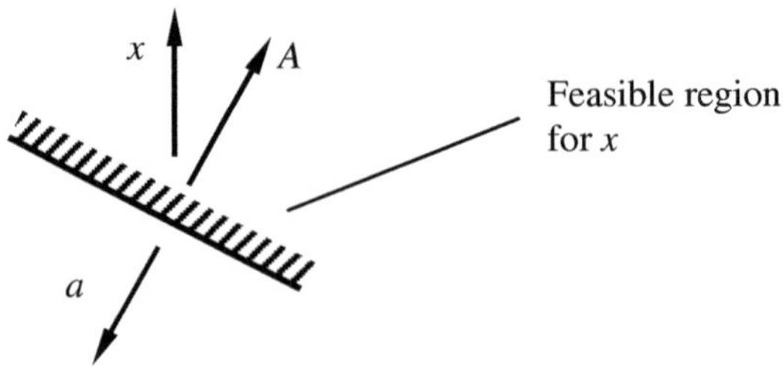

Fig. 2.1. Case $m = 1$

When m=2, the system can be put into the form

$$a + [\, A_1\ A_2 \,] \begin{bmatrix} \lambda_1 \\ \lambda_2 \end{bmatrix} = 0 \quad \Leftrightarrow \quad a^T x < 0 \text{ and } \begin{bmatrix} A_1 \\ A_2 \end{bmatrix} x \geq 0\lambda$$

In Fig. 2.2 this system is given geometric interpretation by regarding the rows of A to be n-vectors. Quite briefly, this figure attempts to show that if $-a$ can be expressed as a positive linear (convex) combination of A_1 and A_2 then there is no feasible x which corresponds to a decreasing value of the objective function.

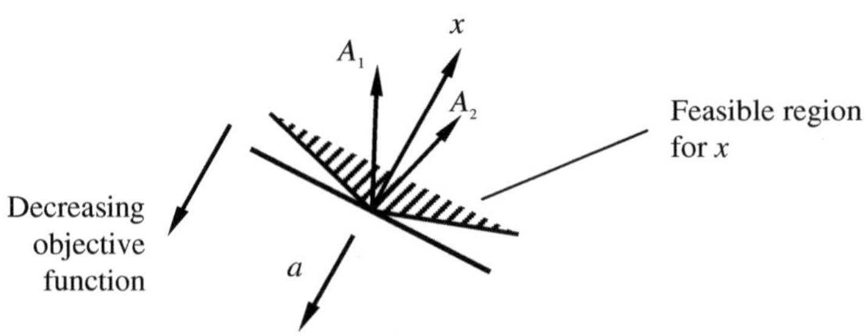

Fig. 2.2. Case $m = 2$

For an algebraic proof of Farkas' Lemma, the reader should consult Simonnard (1966).

With regard to the Kuhn–Tucker theorem, the statement $\lambda_i\, g_i = 0$ is added to handle the case in which some constraints may not be *tight.* In this case the variation dx is not required to lie within the corresponding tangent planes of these constraint surfaces and the associated Lagrange multipliers are taken to be zero.

As an example, consider the problem of Luenberger (1984, p. 315):

$$\textit{minimize}\ 2x^2 + 2xy + y^2 - 10x - 10y$$

$$\textit{subject to}\ x^2 + y^2 \le 5 \quad \textit{and} \quad 3x + y \le 6$$

The Kuhn–Tucker conditions for this problem are

$$\nabla f + \lambda^T \nabla g = 0 \qquad \lambda^T g = 0 \qquad \lambda \ge 0$$

or

$$\begin{aligned} 4x + 2y - 10 + \lambda_1(2x) + \lambda_2(3) &= 0 \\ 2x + 2y - 10 + \lambda_1(2y) + \lambda_2(1) &= 0 \\ \lambda_1\left(x^2 + y^2 - 5\right) &= 0 \\ \lambda_2(3x + y - 6) &= 0 \\ \lambda_1 &\ge 0 \\ \lambda_2 &\ge 0 \end{aligned}$$

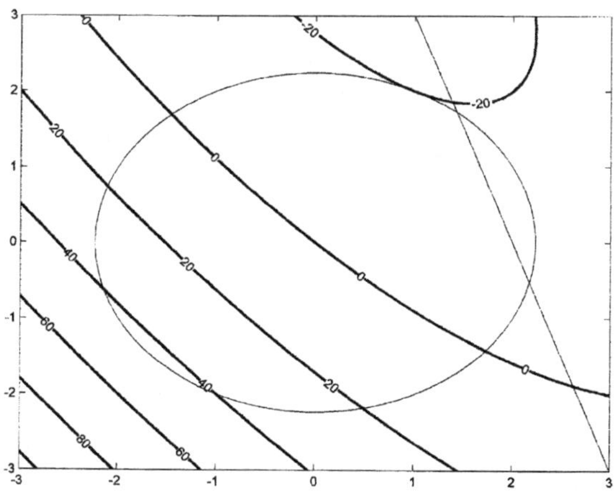

Fig. 2.3. Optimization Example

Rather than solving this system directly, it can be verified that there is a solution with x=1, y=2, λ_1 = 1, and λ_2 = 0. This corresponds to the first constraint being tight and the second not tight. (See Fig. 2.3. Optimization in this case attempts to move downhill, up and to the right, while staying inside the circle and to the left of the line.) Note that the Kuhn–Tucker conditions are satisfied at this point.

2.3 Calculus of Variations

When dealing with continuous systems, it is common to encounter integrals that are to be maximized or minimized subject to certain conditions. This is the realm of calculus of variations.

First of all there is the *fundamental lemma of calculus of variations*, which states that

$$\int_{x_0}^{x_1} \eta(x)\varphi(x)dx = 0 \quad \textit{for all functions } \eta(x) \quad \Rightarrow \quad \varphi(x) = 0$$

Proof: If the function $\eta(x)$ is made sufficiently discontinuous (a delta function, for example) the proof becomes obvious. See Courant and Hilbert (1953) for another discussion of this lemma.

Calculus of variations deals with the following problem (Fig. 2.4):

$$\min_{y} \quad I = \int_{x_0}^{x_1} f(x, y, y', y'')dx$$

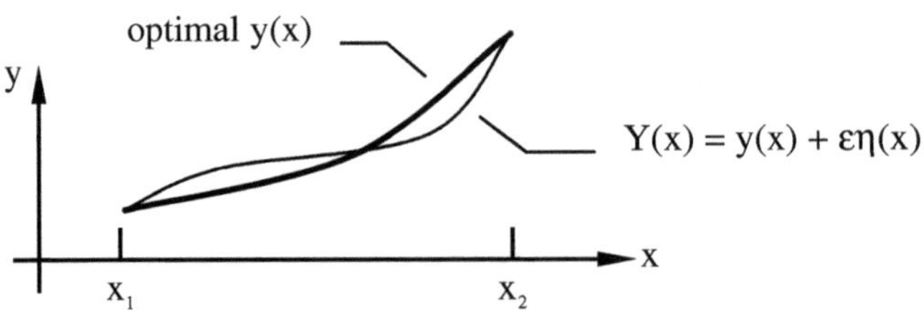

Fig. 2.4. Variation of y(x)

The technique used to solve this problem is to write the integral as a function of a single parameter ε as

$$I(\varepsilon) = \int_{x_1}^{x_2} f(x, Y, Y', Y'')dx$$

with

$$Y = y + \varepsilon\eta(x) \qquad \partial Y/\partial\varepsilon = \eta$$
$$Y' = y' + \varepsilon\eta'(x) \qquad \partial Y'/\partial\varepsilon = \eta'$$

The optimal *I* then satisfies

$dI/d\varepsilon = 0$ at $\varepsilon = 0$

But

$$dI/d\varepsilon\big|_{\varepsilon=0} = \int_{x_1}^{x_2} \left(\frac{\partial f}{\partial y}\eta + \frac{\partial f}{\partial y'}\eta + \frac{\partial f}{\partial y''}\eta'' \right) dx = 0$$

Integrating by parts gives

$$I'(0) = \int_{x_1}^{x_2} \left(\frac{\partial f}{\partial y} - \frac{d}{dx}\frac{\partial f}{\partial y'} + \frac{d^2}{dx^2}\frac{\partial f}{\partial y''} \right) \eta\, dx + boundary\, terms = 0$$

Using the lemma, Euler's equation follows directly as

$$\frac{\partial f}{\partial y} - \frac{d}{dx}\frac{\partial f}{\partial y'} + \frac{d^2}{dx^2}\frac{\partial f}{\partial y''} = 0$$

For a simple example of the use of Euler's equation, return to the beam problem from Chapter 1,

Find m to minimize $\int |m|\, dx$ subject to $m'' = -w$

Using Lagrange multipliers, form the Lagrangian

$$L = \int (|m| + \lambda (m'' + w))dx$$

Applying Euler's theorem directly gives the optimality condition

$\text{sgn}\, m + \lambda'' = 0$

We will return to the application of this optimality condition in Chapter 3.

2.4 Newton's Method

Given a vector function (an *n*-vector) **F(x)**, it is possible to *linearize* this function at some point $\mathbf{x}_0$ by making a Taylor series expansion and neglecting terms of order greater than 1 as

$$F(x) \cong F(x_0) + \nabla F(x_0) \cdot (x - x_0)$$

or simply

$$F(x) \cong F_0 + \nabla F_0 \cdot dx$$

Using Newton's method to solve a nonlinear system of n equations in n unknowns then requires iterating, at each step solving a system of linear equations,

$$F(x) = 0 \quad \Rightarrow \quad F(x) \cong F_0 + \nabla F_0 \cdot dx = 0 \quad \Rightarrow \quad dx = -(\nabla F_0)^{-1} F_0$$

The case (see below) for a single equation in a single variable x is most familiar.

Newton's method has the remarkable property of quadratic convergence, which will now be demonstrated for a scalar system. For a vector system the reader is referred to Isaacson and Keller (1966). If a function F(x) is expanded in a Taylor series about the point x_0 it follows that (Taylor series with remainder)

$$F(x) = F(x_0) + F'(x_0)(x - x_0) + \tfrac{1}{2} F''(\xi)(x - x_0)^2$$

where ξ is some point in the interval (x, x_0). In Newton's method an improved approximation x_1 is computed using only linear terms as

$$x_1 = x_0 - F(x_0)/F'(x_0)$$

The exact root x^* satisfies the equation

$$0 = F(x_0) + F'(x_0)(x^* - x_0) + \tfrac{1}{2} F''(\xi)(x^* - x_0)^2$$

or

$$0 = F(x_0) / F'(x_0) + (x^* - x_0) + \tfrac{1}{2} F''(\xi)(x^* - x_0)^2 / F'(x_0)$$

Combining equations gives

$$0 = x_0 - x_1 + (x^* - x_0) + ½\ F''(\xi)\,(x^* - x_0)^2 / F'(x_0)$$

finally

$$x_1 - x^* = (x^* - x_0)^2\ ½\ F''(\xi) / F'(x_0)$$

Now if $|½\ F''(\xi) / F'(x_0)| \leq 1$ it follows that

$$|x^* - x_1| \leq (x^* - x_0)^2$$

or that the convergence is quadratic in the neighborhood of the point x^*.

2.4.1 An Example of Newton's Method

Consider a beam with a rectangular section subjected to an axial load *P* and bending moment *M* (see Fig. 2.5). Given that the ratio of the base to the height of the cross section is fixed, we seek the height of the section that will support these forces and satisfy an allowable stress requirement.

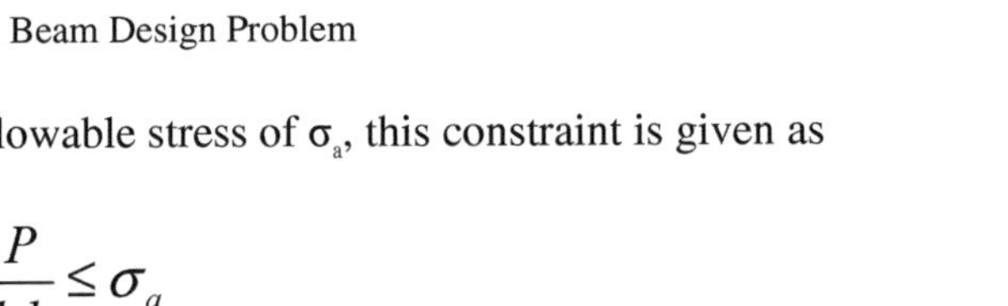

Fig. 2.5. A Beam Design Problem

For an allowable stress of σ_a, this constraint is given as

$$\frac{6M}{bh^2} + \frac{P}{bh} \leq \sigma_a$$

Anticipating optimization in the work that follows, this inequality will be treated as an equality. That is, we will assume that the optimal design will have the smallest value of h for which this allowable stress criterion is satisfied as an equality. Parameters for this example are tabulated below:

Parameter	Value
M	100 k-in.
P	20 k
α	0.3
σ_a	20 ksi

Introducing a parameter for the aspect ratio of the cross section as $\alpha = b/h$ the inequality can be written as

$$F(h) = 6M + hP - \alpha h^3 \sigma_a = 0$$

Here F is simply a cubic equation for h that will be solved as an equality using Newton's method. The required first derivative in this case is

$$F' = P - 3\alpha h^2 \sigma_a$$

Starting with an arbitrary initial value of $h = 10$ in., the first five iterations are

Table 2.1 Iterations of Newton's Method

Step	h (in.)	F	F'	dh
1	10	−5200	−1780	−2.92
2	7.08	−1387	−882	−1.57
3	5.51	−292	−526	−0.55
4	4.95	−29	−421	−0.07
5	4.88	0	−409	0.00

For this example, five iterations give a solution that is accurate to within two-tenths of an inch. One may also examine the solution space of this example graphically by plotting $F(h)$ (Fig. 2.6).

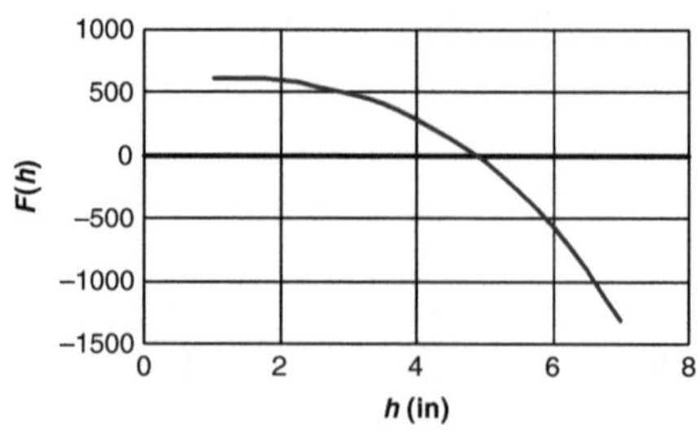

Fig. 2.6. Newton's Method

Referring to the equations of Newton's method for a single variable given above, one may now consider a graphical interpretation of a single step from x_i to x_{i+1} for a function $y = F(x)$. (see Fig. 2.7).

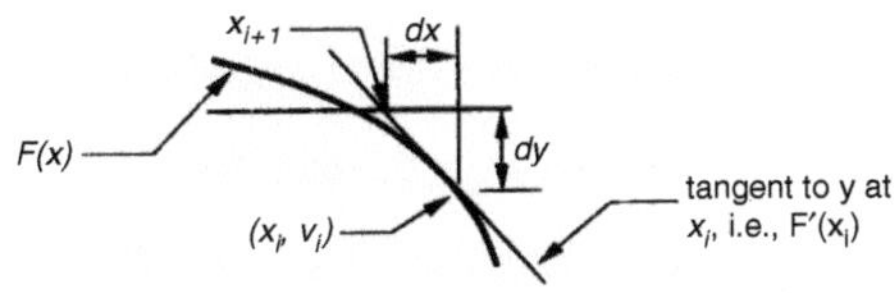

Fig. 2.7. A Typical Step of Newton's Method

Rearranging the identity $F' = dy/dx$ gives $dx = dy/F'$. In seeking a root $F(x) = 0$, dy is seen graphically to be equal to the magnitude of the function at x_i, noted here as $F(x_i)$. Newton's method is thus derived graphically as $x_{i+1} = x_i - F(x_i)/F'(x_i)$.

It can be instructive to examine some details of this example:

- One must start with an initial value. For this example the initial value was 10 in. although any non-zero value could be used because of the nature of the problem to be solved. It will be seen later that sometimes obtaining an initial value is non-trivial. Further, the solution may depend on the initial value if several solutions exist.

- The procedure is iterative and incremental. Throughout this book nonlinear problems will be solved using an iterative approach. In general, increments are sought which smoothly move toward the solution.

- Each step of Newton's method is linear. Each step (i.e., computing dh) is made using the tangent of the function (i.e., a straight line) at a given point. The function itself is, of course, nonlinear but a linear approximation made at the running solution h. As indicated in this example, the incrementally linear approach will often give excellent convergence.

- The transition from inequality to equality of the function F may be considered as what will be termed later a *tight* constraint. Tight constraints are defined as those that govern the solution. This means that although there may be several constraints, all of them may not be active. Generally it is not possible to determine which constraints are tight before a solution is developed.

- The solution is approximate. From a practical point of view one may consider the solution exact since the error may be made arbitrarily small by increasing the number of iterations. However, one should still keep in mind that strictly speaking, the solution is approximate, as will normally be the case with any iterative solution.

- All solutions listed in Table 2.1 are within the *feasible space*, meaning that they all satisfy the constraint. This is not a general property of Newton's method.

2.5 Linear Programming

Historically, the workhorse tool of optimization has been linear programming (Damkilde et al., 1994). While more general tools have struggled with their applications, linear programming has continued to produce results for real (large) systems. In this section an introduction to linear programming will be sketched. Linear programming solvers are discussed in detail in Appendix A.

Linear programming is, of course, mathematical programming where both the objective function and the constraints are linear. The standard form of linear programming seems to be

Find x to minimize $c^T x$ subject to $Ax = b$, $x \geq 0$

The so-called *simplex method* is frequently used to solve linear programming problems. It is based upon the following procedure. Partition the matrix A after some possible rearrangement of rows and columns as $A = [\, A_B \, , A_N \,]$ so that A_B is nonsingular. (Note that a partition of A implies a partition of x.) It follows that

$$Ax = b \quad \Rightarrow \quad A_B x_B + A_N x_N = b \quad \Rightarrow \quad x_B = A_B^{-1}(b - A_N x_N)$$

This implies that

$$\begin{aligned} c^T x &= c_B^T x_B + c_N^T x_N = c_B^T A_B^{-1}(b - A_N x_N) + c_N^T x_N \\ &= c_B^T A_B^{-1} b + (c_N^T - c_B^T A_B^{-1} A_N) x_N \end{aligned}$$

Note that this decomposition of the matrices A and x has allowed the objective function to be written in terms of the variables x_N. This decomposition is fundamental to the *simplex method.* The variables x_B and x_N are called basic and non-basic variables, respectively. The simplex method starts with some set $x_N = 0$. (Starting may be nontrivial.) The coefficient of each $(x_N)_i$ in the expression $c^T x = 0$ is then examined. If any coefficient is negative, that variable can be increased to reduce the objective function. The limit of this increase is determined by some element of x_B going to zero. By this process there is one variable going into the basis and one variable coming out. This procedure gives rise to a theorem of linear programming which states that "if there exists an optimal solution, there exists a basic optimal solution".

2.5.1 An Example of the Simplex Method

The simplex method will now be discussed by way of an example. Consider the following problem:

$$\begin{aligned} &\textit{maximize} \quad \phi = 5x_1 + 4x_2 + 3x_3 \\ &\textit{subject to} \quad 2x_1 + 3x_2 + x_3 \le 5 \\ &\qquad 4x_1 + x_2 + 2x_3 \le 11 \\ &\qquad 3x_1 + 4x_2 + 2x_3 \le 8 \\ &\qquad x_1, x_2, x_3 \ge 0 \end{aligned}$$

Slack variables are first added to convert this problem to one with *equality* constraints as

$$\begin{bmatrix} 2 & 3 & 1 & 1 & & \\ 4 & 1 & 2 & & 1 & \\ 3 & 4 & 2 & & & 1 \end{bmatrix} \begin{bmatrix} x_1 \\ x_2 \\ x_3 \\ z_1 \\ z_2 \\ z_3 \end{bmatrix} = \begin{bmatrix} 5 \\ 11 \\ 8 \end{bmatrix}$$

In this case the starting solution is obvious: $x = [0\ \ 0\ \ 0\ \ 5\ \ 11\ \ 8]$

Initial basis: 4,5,6

x_1 into basis

$$\begin{aligned} 2x_1 + z_1 &= 5 \quad \Rightarrow \quad & x_1 &= 5/2 \quad (\textit{smallest}) \\ 4x_1 + z_2 &= 11 & x_1 &= 11/4 \\ 3x_1 + z_3 &= 8 & x_1 &= 8/3 \end{aligned}$$

$\therefore z_1$ *out of basis*

New basis

$$\begin{bmatrix} 1 & 1.5 & 0.5 & 0.5 & & \\ 0 & -5 & 0 & -2 & 1 & \\ 0 & -0.5 & 0.5 & -1.5 & & 1 \end{bmatrix} \begin{bmatrix} x_1 \\ x_2 \\ x_3 \\ z_1 \\ z_2 \\ z_3 \end{bmatrix} = \begin{bmatrix} 2.5 \\ 1 \\ 0.5 \end{bmatrix}$$

New objective function

$$x_1 = 2.5 - 1.5x_2 - 0.5x_3 - 0.5z_1 \quad \Rightarrow \quad \phi = 12.5 - 3.5x_2 + 0.5x_3 - 2.5z_1$$

x_3 into basis

$$x_1 + 0.5x_3 = 2.5 \quad \Rightarrow \quad x_3 = 5$$
$$z_3 + 0.5x_3 = 0.5 \quad \Rightarrow \quad x_3 = 1 \quad (smallest)$$

$$\therefore z_3 \quad \textit{out of basis}$$

New basis

$$\begin{bmatrix} 1 & 2 & 0 & 2 & & -1 \\ 0 & -5 & 0 & -2 & 1 & \\ 0 & -1 & 1 & -3 & & 2 \end{bmatrix} \begin{bmatrix} x_1 \\ x_2 \\ x_3 \\ z_1 \\ z_2 \\ z_3 \end{bmatrix} = \begin{bmatrix} 2 \\ 1 \\ 1 \end{bmatrix}$$

New objective function

$$x_3 = 1 + x_2 + 3z_1 - 2z_3 \quad \Rightarrow \quad \phi = 13 - 3x_2 - z_1 - z_3$$

At this point there is no possible improvement. The final solution is $x_2 = z_1 = z_3 = 0$, $x_1 = 2$, $z_2 = 1$, $x_3 = 1$.

Program 30 on the CD discusses the simplex method and this example.

2.5.2 Interior Point Methods

The simplex method, the workhorse method of linear programming, is said to be an *exterior* method since it works on the exterior of the feasible region. Following the publication, in 1984, of a paper by Karmarkar, there was a flurry of activity over what have been called *interior* methods. Roughly, the idea is that it may be more efficient, when solving a linear programming problem, to move through the interior to find an optimal point rather than simply moving around the boundary as the simplex method does.

This section will give the reader a brief look at interior methods. For more information the reader is referred to the excellent book by Arbel (1993) and the more comprehensive work of Wright (1997).

The so-called *primal* method starts with the linear programming problem in the form

$$\textit{minimize } c^T x \quad \textit{subject to } Ax = b \textit{ and } x \geq 0$$

and looks for an incremental change dx, such that

$$c^T x_{new} \leq c^T x_o \quad \textit{and} \quad Ax_{new} = b$$

This implies that

$$c^T dx \leq 0 \quad \textit{and} \quad Adx = 0$$

It turns out that if dx is taken to be $-Pc$, with the *projection* $P = I - A^T(AA^T)^{-1}A$ then

$$c^T dx = -c^T Pc = -c^T P^2 c = -\|Pc\| \leq 0$$

since $P = P^T$ and $P^2 = P$.

There are two other points to be made. First, it is effective to work with variables that are *scaled*. This is done by transforming the problem as $x_1 = D^{-1}x$ with the matrix D diagonal having its nonzero values as the unscaled variables. Finally the increment dx is scaled so that x always remains within the feasible region. Some of the code outlined by Arbel is given in Programs 31 and 32 on the CD.

2.6 Sequential Linear Programming

Given an optimization problem such as minimize $f(x)$ subject to $\mathbf{g}(\boldsymbol{x}) \leq \mathbf{0}$, there is a *Newtonian* approach in which the system is linearized at some point x_0 and the simplified system is solved as

$$\text{minimize } f(x_0) + \nabla f(x_0) \cdot dx \qquad \text{subject to} \qquad \mathbf{g}(\mathbf{x}_0) + \nabla \mathbf{g}(x_0) \cdot \mathbf{dx} \leq \mathbf{0}$$

This problem results in a new approximate solution as $x \Rightarrow x + dx$.

Repeated application of this step results in what is called *sequential linear programming* (El Hallabi 1994) since each step involves solving a linear programming problem. Since linear programming has the ability of dealing efficiently with large systems, sequential linear programming has considerable potential as a tool for structural optimization. (It should be noted here that sequential linear programming commonly requires the application of move limits. That point will be discussed in Chapter 3.)

It turns out that it is an easy matter to code this algorithm (see below). In the case described here the problem is to

$$\begin{aligned} \textit{minimize} \quad & f = -x_1^3 - 2x_2^2 + 10x_1 - 6 - 2x_2^3 \\ \textit{subject to} \quad & g_1 = 10 - x_1 x_2 \geq 0 \\ & g_2 = x_1 \geq 0 \\ & g_3 = 10 - x_2 \geq 0 \end{aligned}$$

There are several steps in this program:

- The objective function and the constraints must be described for the algorithm. That is done in the subroutine ROUTINES. (For each new problem this subroutine has to be replaced with an appropriate subroutine.)
- In the main program x, n, and m describe the starting values, the size of the x-vector, and the number of constraints.
- DO 4 ITER=1,NIT describes the iterative process. The linear programming solver DLPRS from the IMSL package is called at each iteration.
- Note the upper and lower bounds are specified as 10% of the running value of the variable under consideration.

Program03

```
c       SEQUENTIAL LP SOLVER
        USE IMSL
        DIMENSION X(50),G(50),DELG(50,50),B(50),DELF(50)
       & ,IRTYPE(50),XLB(50),XUB(50),XSOL(50),DSOL(50)
       & ,A(50,50)
        X(1)=1.
        X(2)=1.
```

```
      N=2
      M=3
      LDA=50
      NIT=100
      CFAC=.1
      DO 4 ITER=1,NIT
      CALL ROUTINES (X,F,G,DELG,N,ITER,DELF)
      DO 1 I=1,M
      B(I)=-G(I)
      IRTYPE(I)=2
    1 CONTINUE
      DO 2 I=1,N
      XLB(I)=-CFAC*X(I)
    2 XUB(I)= CFAC*X(I)
      CALL DLPRS(M,N,DELG,LDA,B,B,DELF,IRTYPE,XLB,XUB,OBJ,
     & XSOL,DSOL)
      DO 3 I=1,N
    3 X(I)=X(I)+XSOL(I)
    4 CONTINUE
      STOP
      END
C
      SUBROUTINE ROUTINES (X,F,G,DELG,N,ITER,DELF)
      DIMENSION X(50),G(50),DELG(50,50),DELF(50)
      F=-(X(1)**3)-2.*(X(2)**2)+10.*X(1)-6-2.*(X(2)**3)
      DELF(1)=-3.*(X(1)**2)+10.
      DELF(2)=-4.*X(2)-6.*(X(2)**2)
      G(1)=10.-X(1)*X(2)
      G(2)=X(1)
      G(3)=10.-X(2)
      DELG(1,1)=-X(2)
      DELG(1,2)=-X(1)
      DELG(2,1)=1.
      DELG(2,2)=0.
      DELG(3,1)=0.
      DELG(3,2)=-1.
      WRITE(60,*)ITER,F,(X(I),I=1,N)
      WRITE(61,*)ITER,F
      RETURN
      END
```

The performance is rather good in this case but this algorithm does not always work.

Output from Program 3

Iteration	Objective Function	x_1	x_2
1	-1.000000	1.000000	1.000000
2	-2.811000	9.000000E-01	1.100000
3	-4.902763	8.100000E-01	1.210000
4	-7.356440	7.290000E-01	1.331000
5	-10.285466	6.561000E-01	1.464100
6	-13.842975	5.904900E-01	1.610510
7	-18.232380	5.314410E-01	1.771561
8	-23.721951	4.782969E-01	1.948717
9	-30.664513	4.304672E-01	2.143589
10	-39.523773	3.874205E-01	2.357948
11	-50.909420	3.486784E-01	2.593743

12	-65.623672	3.138106E-01	2.853117
13	-84.723083	2.824295E-01	3.138429
14	-109.600487	2.541866E-01	3.452271
15	-142.093719	2.287679E-01	3.797499
16	-184.629623	2.058911E-01	4.177248
17	-240.415451	1.853020E-01	4.594974
18	-313.692261	1.667718E-01	5.054471
19	-410.071808	1.500946E-01	5.559918
20	-536.983582	1.350852E-01	6.115910
21	-704.267944	1.215767E-01	6.727500
22	-924.964844	1.094190E-01	7.400250
23	-1216.360352	9.847709E-02	8.140276
24	-1601.377686	8.862938E-02	8.954304
25	-2110.426025	7.976644E-02	9.849734
26	-2205.282471	7.178980E-02	10.000000
27	-2205.354248	6.461082E-02	10.000000
98	-2205.999756	3.643540E-05	10.000000
99	-2205.999756	3.279186E-05	10.000000
100	-2205.999756	2.951267E-05	10.000000

The disk includes an EXCEL Solver solution of this problem as Program04.

2.7 Other Methods of Mathematical Programming

There is a world of mathematical programming algorithms out there (see, for example, Ecker and Kupferschmid 1991 and Kumar, 2000). In particular, Ecker and Kupferschmid describe an "ellipsoidal algorithm" that is quite robust. Some of these will be returned to later in this text. For problems of structural optimization the reader may wish to consult Schittkowski et al., 1994. (Schittkowski has, over the years, made available computer code for structural optimization.)

Some of the more commonly used algorithms include sequential quadratic programming.

Since the Excel Solver program (Fylstra et al. 1998) is given a lot of use in this text, we will refer briefly to the generalized reduced gradient method of solving mathematical programming problems about which Solver is built. When discussing general approaches to solving mathematical programming problems, it is worthwhile to start with the fact that simple gradient methods just do not work. That is, if you try to move in the direction of the gradient of the objective function it is the conventional wisdom that you get hung up on the constraint surfaces.

Briefly, the generalized reduced gradient (GRG) method works something like this. Rather than working with the typical nonlinear programming formulation

minimize $f(\mathbf{x})$ subject to $\mathbf{g}(\mathbf{x}) \leq 0$

it is convenient in this case to start with a formulation of minimize f(**x**) subject to **g**(**x**) = 0

This can be done on the basis of introducing slack variables or even just working with the tight constraints and ignoring the others. Now the system is linearized and the elements of **x** partitioned after possibly reordering them as

$$\min f_0 + \nabla f_0 dx \qquad \textit{subject to} \quad g_0 + \nabla g_0 dx = 0$$

or

$$\min f_0 + \nabla f_{01} dx_1 + \nabla f_{02} dx_2 \quad \textit{subject to} \quad g_0 + \nabla g_{01} dx_1 + \nabla g_{02} dx_2 = 0$$

Assuming that ∇g_{02} has the appropriate rank, it follows that

$$dx_2 = -\nabla g_{02}{}^{-1} \nabla g_{01} dx_1$$

and

$$\begin{aligned} \nabla f\, dx &= \nabla f_1 dx_1 + \nabla f_2 dx_2 \\ &= (\nabla f_1 - \nabla f_2 \nabla g_{02}{}^{-1} \nabla g_{01}) dx_1 \end{aligned}$$

This is the reduced gradient. A search can now be carried out in the direction of the reduced gradient projected on the working surface $\nabla g = 0$.

2.8 Genetic Algorithms

Genetic algorithms (GA) are a member of a class of *heuristic* algorithms that represent an active research area today (see, for example, Sivakumar et al. 2004). Broadly speaking, they model the natural selection process of biology. This section briefly describes some of the features of the GA algorithm but we leave it to the reader to pursue this area more fully. Genetic algorithms offer their users the potential of creative behavior and, in that sense, are part of an exciting field. They can be, at the same time, quite arbitrary and tend to neglect gradients, which the authors see as a serious shortcoming. But taken at their most general, it cannot be said for certain that they leave anything out.

In the following we will describe some features of a very simple genetic algorithm. In this case a design (chromosome) is represented by a binary, finite string. This string contains a description of the elements of the design (genes) that are under study. There must also be a metric (fitness) that can be used to determine that one design is better than another. In the paper cited above, one of their examples is a 10-bar truss whose design elements are the bar areas. The fitness in this case is the weight of the structure.

The question then is how to generate new designs given a starting design. Classically this involves *crossover* and *mutation.* With crossover, two designs are compared and parts taken from each to form a new design. With mutation, there is the potential for change within an individual design. For a sizeable design represented by a binary string, there are a large number of designs possible. It is this large number that creates problems for genetic algorithms. On the other hand, the algorithms used in this text work in the small using, for example, Taylor series methods. And if you work in the small you will not usually get radically different designs. This is what drives the designer to use heuristic methods.

2.9 Problems

1. Use the simplex method to solve the following problem:

$$\text{maximize } 6x_1 + 8x_2 + 5x_3 + 9x_4$$

subject to

$$2x_1 + x_2 + x_3 + 3x_4 \le 5$$
$$x_1 + 3x_2 + x_3 + 2x_4 \le 3$$
$$x_1, x_2, x_3, x_4 \ge 0$$

2. Discuss the solution of (Ecker and Kupferschmid 1991, pp. 315)

$$\text{minimize} \quad (x_1 - 20)^4 + (x_2 - 12)^4$$

$$\text{subject to} \quad 8\,e^{(x_1 - 12)/9} - x_2 + 4 \le 0$$
$$6\,(x_1 - 12)^2 + 25\,x_2 - 600 \le 0$$
$$-x_1 + 12 \le 0$$

(They solve this problem using the so-called ellipsoid algorithm. Try to solve it using sequential linear programming.)

3. Discuss the paper (Koumousis and Georgiou 1994) "Genetic Algorithms in Discrete Optimization of Steel Truss Roofs".

4. Discuss the paper (Fylstra et al. 1998) "Design and Use of the Microsoft Excel Solver".

5. Discuss the use of *barrier* methods (Wright 1997) in linear programming.

6. Discuss the "Optimization Overview" in the Matlab Optimization Toolbox.

3 Sequential Linear Programming and the Incremental Equations of Structures

This chapter presents a general approach to problems of structural optimization. The idea is that when used with the incremental equations of structures, sequential linear programming offers a robust tool for solving structural optimization problems. This is particularly true given the potential for solving large linear programming problems cited in the Preface. Truss and frame problems, including column buckling and shape optimization, are discussed here.

3.1 Introduction

This chapter represents the principal contribution of this text. It argues that linear programming, together with the incremental equations of structures, provides a robust format from which to solve problems of structural optimization. Generally speaking, this approach does not tell you a lot about structures. For example, optimality criteria methods, in some cases, give up statically determinate solutions and information that can be important. The advantage of the methods of this chapter is that they allow you to formulate and solve very complex problems simply. We think, simply enough to place structural optimization alongside the finite element method on the engineer's desk.

Most optimization methods are incremental, typically involving some type of Taylor's series approach. And linear programming is the oldest method used for solving optimization problems. What we hope to show the reader is that linear programming combined with the incremental equations of structures can be useful in structural optimization. Unfortunately, as the devil is in the details, to appreciate this usefulness fully, the reader will have to get involved with the coding of the problems discussed. We have done our best to help the reader with this point.

3.2 The Incremental Equations of Structures

With the *incremental* approach, you start from some initial solution and look for small changes in parameters that will improve the performance of a given system. For example, you might start with some truss design problem and look for changes in bar areas, dA_i , that would reduce stresses or displacements while minimizing the construction cost. This approach has the advantage of *starting*

W.R. Spillers, K.M. MacBain, *Structural Optimization*,
DOI 10.1007/978-0-387-95865-1_3,

with a linearized problem. This linearized optimization problem is, of course, a linear programming problem.

The incremental equations of structures (see Table 3.1) work in the spirit of Newton's method. We ask what the effect of changing the member stiffness matrix K to K+dK is. We find that $d\delta = -K_E^{-1}N^T\, dK\, \Delta$. (See Appendix B for a detailed discussion of the truss problem.)

Table 3.1 The Equations of Structures

	Equations of structures	Incremental form
Equilibrium equations	$N^T F = P$	$N^T dF = 0$
Constitutive equations	$F = K\,\Delta$	$dF = dK\,\Delta + K\,d\Delta$
Node/member displacement	$\Delta = N\,\delta$	$d\Delta = N\,d\delta$

Here

- F, Δ – Member force and displacement matrices
- P, δ – Joint force and displacement matrices
- K – Primitive stiffness matrix (for trusses, K is diagonal with $K_{i,i} = A_i E / L_i$)
- N – Generalized incidence matrix

For trusses, the simplest case, A_i and L_i are the bar area and length, respectively, and E is Young's modulus. The lower case "d" is used to indicate a (small) change in a variable. For example, in truss problems a joint displacement δ or a change in displacement $d\delta$ produces a change in the joint coordinate matrix R as $d\delta \Rightarrow R \rightarrow R + dR = R + d\delta$.

The displacement method of structures is usually solved as

$$\delta = \left(N^T K N\right)^{-1} P \quad \rightarrow \quad \Delta = N\delta \quad \rightarrow \quad F = K\Delta$$

The incremental version of this process is

$$d\delta = -\left(N^T K N\right)^{-1} N^T dK\Delta \quad \rightarrow \quad d\Delta = Nd\delta \quad \rightarrow \quad dF = dK\Delta + Kd\Delta$$

Here a change in stiffness dK produces a change in joint displacement $d\delta$, etc., and note that the term $dK\ \Delta$ is a column matrix and can be thought of as containing "scaled" values of dA_i and dI_i later when discussing frames.

3.3 Application to Structural Optimization

Typically, a structural optimization problem is stated as find K, the matrix of member stiffnesses, to satisfy the equations of structures, together with some displacement or stress constraints and minimize the structural volume or cost. The incremental version of this problem starts with some given solution and looks for a dK that satisfies certain constraints. This incremental problem is then a linear programming problem. Some details of the application of the incremental approach are now discussed:

- Move limits: Since the linearized problem may imply solutions in which dK may go to infinity, move limits are generally required. (Roughly speaking, the incremental problem may want to go to infinity but it can still provide a useful *direction* for dK.) Move limits must be dealt with carefully as discussed in the recent work of Lamberti and Pappalettere (2000, 2004). Conventional wisdom going back many years has move limits of 10–15% used commonly.
- Scaling: The authors have found that scaling variables can be useful. For example, in the classic 25-bar truss problem (Wang et al., 2002) discussed below, with stress and displacement constraints, we found it useful to scale to a point at which the displacement constraints are satisfied and then look for an incremental solution. (With regard to scaling, clearly $K \rightarrow \alpha K \Rightarrow \delta \rightarrow \delta / \alpha$ when α is a scalar parameter.) That is, if you know that a problem like this is driven by a displacement constraint, it may be efficient to take care of that fact early in the solution of the problem.
- Regions of trust: While there may be many constraints in a problem to be dealt with, it is common that all the constraints may not be *active*. When dealing with the incremental approach it can be useful to establish *regions of trust* in which tight constraints are identified and other constraints are temporarily neglected. In this case you do not know the size of the linear programming problem to be solved until that particular step in the optimization problem is executed.
- The objective function: In the case of truss problem cited above, the volume is simply the sum of the product of each bar area times its length. In the incremental version, the objective function is then the sum of the product of each *incremental* bar area, $dA_{i,}$ and its length. This is easily related to $(dK)_{i,i} = dA_i \, E / L_i$. In other types of structures similar steps must be taken.

- Displacement constraints: For a given displacement constraint, say $\delta_I + d\delta_I \le c$, there is an implied inequality saying
$$d\delta_I = -\sum_j \left[\left(N^T K N \right)^{-1} N^T \right]_{I,j} (dK\Delta)_j \le c - \delta_i$$
- Stress constraints: Stress constraints can be handled like displacement constraints but it is worthy of note that this is most easily done through the incremental member displacement $d\Delta$ since, for the truss again, $d\Delta_I = d\sigma_I L_I / E$. These would appear as

$$\sigma_I + d\sigma_I \le c \quad \Rightarrow \quad d\Delta_I = d\sigma_I L_I / E \le (c-\sigma_I) L_I / E \quad \Rightarrow$$

$$(Nd\delta)_I \le (c-\sigma_I) L_I / E \quad \Rightarrow$$

$$\left(N\left\{-(N^T K N)^{-1} N^T dK\Delta\right\}\right)_I \le (c-\sigma_I) L_I / E \quad \Rightarrow$$

$$-\sum_j \left\{N(N^T K N)^{-1} N^T\right\}_{I,j} (dK\Delta)_j \le (c-\sigma_I) L_I / E$$

3.4 An Example with a Displacement Constraint

Figure 3.1 shows a truss design problem. In this case the only constraint is a displacement constraint of 1.3″ at the top, right node. This is the most simple application possible of the use of sequential linear programming and the incremental equations of structures. It is Program 9 on the CD.

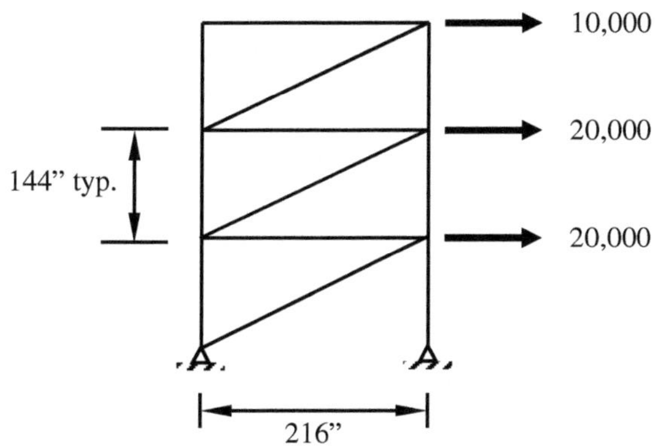

Fig. 3.1. Truss Design

More formally, the problem is simply, find the bar areas A_i to minimize the volume of material, $\Sigma\ A_i\ L_i$, while keeping the displacement at the top right-hand node less than 1.3″. This problem will be returned to later in this book when displacements are again considered.

In order to understand this computer program which is built around the two-dimensional truss problem, the reader might find it useful to go over the two-dimensional truss analysis computer program associated with Appendix B. Program 9 has several steps:

- Input: There are the usual node coordinates, loads, constraints, etc.
- Call to the subroutine Anal: Anal is a typical truss analysis program. It is called initially because the incremental solution needs forces and displacements to get started.
- Call to the subroutine OPT: Opt begins by computing the generalized incidence matrix N as the array AN, the matrix K_E as the array C, and the product $-K_E^{-1}N^T$ as the array PROD. This latter array is required by the incremental equations when $d\delta$ is needed. The member areas are also scaled so that the displacement constraint is initially tight.
- Setting up the sequential LP problem. The code above sets up the linear programming problem

$$\text{minimize}\ \Sigma\ c^T x \quad \text{subject to} \quad A\,x \le b$$

that is solved using the IMSL routine DLPRS that is embedded in the FORTRAN compiler. In this problem

x – member areas (s in the code)
c – array of member lengths (f in the code)
A – matrix of coefficients (alp in the code)

There are two constraint cases to be considered:

$$\delta_i \geq 0 \quad \rightarrow \quad \delta_i + d\delta_i \leq const$$
$$\delta_i \leq 0 \quad \rightarrow \quad \delta_i + d\delta_i \geq const$$

Note the factor of 0.8 used to determine whether a constraint is to be considered active or not.

Some of the code from the subroutine Opt() of Program 9 is listed below.

Code from the subroutine Opt()

```
C      DISPLACEMENT CONSTRAINTS
C
 569   CONTINUE
       DO 131 I=1,NDCONS
       JOINT=NJT(I)
       DO 131 J=1,1
       DO 131 K=1,2
       IF(DCONS(I,K).EQ.0.) GO TO 131
       NCOMP=2*JOINT-2+K
       IF(DEL(J,NCOMP).LT.0.) GO TO 132
       RAT=DABS(DEL(J,NCOMP))/DCONS(I,K)
       IF(RAT.LT..8)GO TO 131
       NROWS=NROWS+1
       DO 134 K1=1,NB
       ALP(NROWS,K1)=PROD(NCOMP,K1)*E*DELTA(J,K1)/ALEN(K1)
  134  CONTINUE
       ITYPE(NROWS)=1
       BU(NROWS)=DCONS(I,K)-DEL(J,NCOMP)
       GO TO 131
  132  CONTINUE

       RAT=DABS(DEL(J,NCOMP))/DCONS(I,K)
       IF(RAT.LT..8)GO TO 131
       NROWS=NROWS+1
       DO 334 K1=1,NB
       ALP(NROWS,K1)=PROD(NCOMP,K1)*E*DELTA(J,K1)/ALEN(K1)
  334  CONTINUE
       ITYPE(NROWS)=2
       BL(NROWS)=-DCONS(I,K)-DEL(J,NCOMP)
 131   CONTINUE

       DO 70 I=1,NB
       F(I)=ALEN(I)
       XLB(I)=-.2*S(I)
  70   XUB(I)=.2*S(I)
```

```
      WRITE(60,*)' *** NROWS',NROWS
C      WRITE(60,*)((ALP(I,J),J=1,NB),I=1,NROWS)
      CALL DLPRS(NROWS,NB,ALP,MAXC,BL,BU,F,ITYPE,XLB,XUB,OBJ,
     & XSOL,DSOL)
      WRITE(60,61)ITER,OBJ,(I,XSOL(I),S(I),I=1,NB)
   61 FORMAT(I5,E20.8/ (I5,2E20.8))
      VOLUME=0.
      DO 71 I=1,NB
      VOLUME=VOLUME+ALEN(I)*S(I)
   71 S(I)=AMAX1(S(I)+REAL(XSOL(I)),.01)
      WRITE(60,*) '*** VOLUME', VOLUME
 9997 CONTINUE
      RETURN
      END
```

The final bar areas are

Bar	Area
1	0.99999998E-02
2	0.37037036E+00
3	0.14814814E+01
4	0.44444445E+00
5	0.17777778E+01
6	0.40000000E+01
7	0.99999998E-02
8	0.66666669E+00
9	0.20000000E+01
10	0.66769469E+00
11	0.20030839E+01
12	0.33384733E+01

```
*** volume     3302.266846
```

3.5 Adding Stress Constraints

To illustrate the effect of stress constraints, Program 10 takes the above truss problem and adds simple stress constraints. In this case the allowable stress in tension is 18 ksi and the allowable stress in compression is −15ksi. In addition to the steps described above the following are included in this program:

- In order to set up the stress constraints, the matrix $-NK_E^{-1}N^T$ is computed as the array PROD2.
- Additional scaling is used to get started. This involves starting with member areas large enough so that the stress constraints are initially satisfied. (Without doing this we found ourselves setting up an infeasible linear programming problem due to the applied move constraints.)

The additional coding in the subroutine OPT() is shown below. The final bar areas including stress constraints are as follows:

Bar	Area
1	0.99999998E-02
2	0.69551814E+00
3	0.14156625E+01
4	0.69006199E+00
5	0.14675570E+01
6	0.25616677E+01
7	0.99999998E-02
8	0.10265103E+01
9	0.15902245E+01
10	0.10432777E+01
11	0.16227401E+01
12	0.27624009E+01

```
*** volume      2961.618408
```

```
      RAT2=0.
      RATIO=0.
      DO 3 I=1,NB
      STRESS1=FSAVE(1,I)/S(I)
      IF(STRESS1. GT. 0.) GO TO 4
      RAT1=STRESS1/SIGALL(2,I)
      IF(FSAVE(1,I).NE.0.)
     1 S(I)=FSAVE(1,I)/SIGALL(2,I)
      GO TO 5
    4 RAT1=STRESS1/SIGALL(1,I)
      IF(FSAVE(1,I).NE.0.)
     1 S(I)=FSAVE(1,I)/SIGALL(1,I)
```

```
      5 CONTINUE
        RATIO=AMAX1(RATIO,RAT1,RAT2)
        WRITE(60,*) '***RATIOS ',I,RAT1,RAT2,RATIO
       3 CONTINUE
        CALL ANAL()
        DO 31 I=1,NB
        DO 31 J=1,1
        IF(FSAVE(J,I).LT.0.) GO TO 32
        RAT=(FSAVE(J,I)/S(I))/SIGALL(1,I)
        IF(RAT.LT..8)GO TO 31
        NROWS=NROWS+1

        DO 34 K=1,NB

ALP(NROWS,K)=PROD2(I,K)*(E*DELTA(J,K)/ALEN(K))*(E/ALEN(I))
     34 CONTINUE
        ITYPE(NROWS)=1
        BU(NROWS)=SIGALL(1,I)-FSAVE(J,I)/S(I)
        GO TO 31
     32 CONTINUE
        RAT=(FSAVE(J,I)/S(I))/SIGALL(2,I)
        IF(RAT.LT..8)GO TO 31
        NROWS=NROWS+1
        DO 234 K=1,NB

ALP(NROWS,K)=PROD2(I,K)*(E*DELTA(J,K)/ALEN(K))*(E/ALEN(I))
    234 CONTINUE
        ITYPE(NROWS)=2
        BL(NROWS)=SIGALL(2,I)-FSAVE(J,I)/S(I)
   31   CONTINUE
  C
   569  CONTINUE
        DO 70 I=1,NB
        F(I)=ALEN(I)
        XLB(I)=-.2*S(I)
    70  XUB(I)=.2*S(I)
        WRITE(60,*)' *** NROWS',NROWS
        CALL
DLPRS(NROWS,NB,ALP,MAXC,BL,BU,F,ITYPE,XLB,XUB,OBJ,XSOL,DSOL)
        WRITE(60,61)ITER,OBJ,(I,XSOL(I),S(I),I=1,NB)
    61  FORMAT(I5,E20.8/ (I5,2E20.8))
        VOLUME=0.
        DO 71 I=1,NB
        VOLUME=VOLUME+ALEN(I)*S(I)
     71 S(I)=AMAX1(S(I)+REAL(XSOL(I)),.01)
        WRITE(60,*) '*** VOLUME', VOLUME
   9997 CONTINUE
```

3.6 The 25-Bar Truss

This problem (Fig. 3.2) has appeared in many places. It traces back at least to Vanderplaats and Moses (1972). It is something of a proof of concept for the material of this chapter in that it includes many of the features encountered by the practicing engineer. It has both stress and displacement constraints and multiple loading conditions (see Fig. 3.3). In addition to the data in this figure, nodes 1 and 2 are limited to a displacement of 0.35 in. in the x and y directions, the maximum tensile stress for all members is 35 ksi, and the minimum member size is 0.01 in^2. It is Program 11 on the CD.

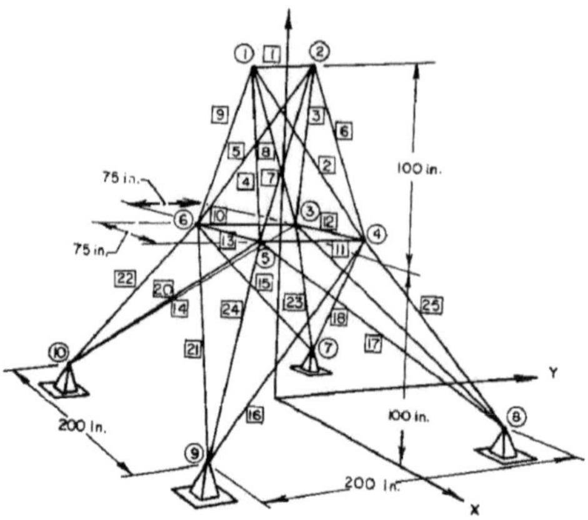

Fig. 3.2. A 25-Bar Truss Problem

Program 11 can be broken down into the following components:

1. Input. There is the usual input in the following order: bars, node supports, load case 1 (coordinates and loads), member connectivity, allowable stresses, displacement constraints, and load case 2.
2. Three primary sections. The code is divided into three elements: TWOLOADS (finds the starting areas using a method discussed later in this text), ANAL (produces a starting solution given the member areas), and OPT (carries out the sequential linear programming).

The progress of iterations is indicated in Fig. 3.4.

(a) Applied loading systems

Load condition	Node	Direction		
		x	y	z
1	1	1000	10,000	−5000
	2	0	10,000	−5000
	3	500	0	0
	6	500	0	0
2	5	0	20,000	−5000
	6	0	−20,000	−5000

(b) Allowable compressive stresses

Member	Allowable stress (lbf/in^2)	Member	Allowable stress (lbf/in^2)
1	−35,092	12, 13	−35,092
2, 3, 4, 5	−11,590	14, 15, 16, 17	−6759
6, 7, 8, 9	−17,305	18, 19, 20, 21	−6969
10, 11	−35,092	22, 23, 24, 25	−11,082

Fig. 3.3. 25-Bar Truss Data

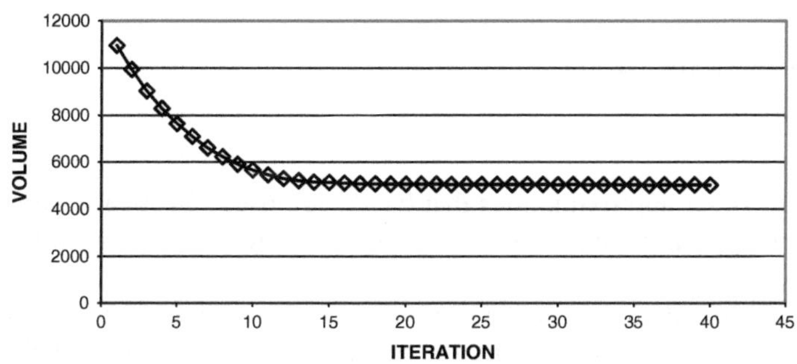

Fig. 3.4. Progress of Iterations

The discussion here will focus on the subroutine OPT. In fact, OPT requires little explanation since it shares much with Program 10. It begins by computing the generalized incidence matrix N, it then checks both the stress constraints and the displacement constraints and scales the solution so that the displacement constraints are tight (this is done arbitrarily to improve convergence of the algorithm), it then computes the matrices $(N^TKN)^{-1}N^T$ and $N(N^TKN)^{-1}N^T$ that are required for the incremental solution, finally displacement and stress constraints are written as described above. From the IMSL package, DLINDS (matrix inversion) and DLPRS (linear programming) are used.

There is also the question of starting values, which seems more like a theoretical rather than a practical issue since the designer typically has some idea of his/her starting design. In this case we tried several starting designs but found nothing to be decisive.

3.7 A Frame Example

Design of a frame is more complicated than the design of a truss for several reasons. First of all there is the question of dimensionality. The plane frame, for example, has three displacement components per joint while the plane truss has only two. But the real problems arise from the fact that the parameters of the frame are simply more difficult. While the truss only deals with the member area, the frame must deal with member area, moment of inertia, and section modulus.

It is common practice to represent the moment of inertia and section modulus as a function of the member area. This is done by curve fitting, for example, the members in the AISC Manual. There is happily considerable literature available to help in this process. In the discussion that follows, for reasons of simplicity, we use the work of Tabak and Wright (1981) who say that using units of inches

$$0 \le I \le 9000$$
$$S = (60.6I + 84100)^{1/2} - 290.$$
$$A = 0.465I^{1/2}$$

There are many other approximations that have been used. Probably the most extensive work has been done by Chan (1997) but the reader may wish to look at Khan (1984) or Choi et al. (1991).

Once this approximation is out of the way, there are other details to take care of. For example, in the plane frame, the member stiffness matrix involves both the member area and the member moment of inertia:

$$K = \begin{bmatrix} K_1 & & 0 & \\ & K_2 & & \\ & & K_3 & \\ 0 & & & \cdots \end{bmatrix} \quad \text{with} \quad K_i = \begin{bmatrix} A_i E / L_i & 0 & 0 \\ 0 & \dfrac{4EI_i}{L_i} & \dfrac{2EI_i}{L_i} \\ 0 & \dfrac{2EI_i}{L_i} & \dfrac{4EI_i}{L_i} \end{bmatrix}$$

Since the incremental equations use the term dK, and if the areas are to be used as variables in the optimization scheme, each A_i and I_i must be replaced by dA_i and dI_i. Using the relationship above, $I = (A/0.465)^2$ which implies that $dI = 2\ A\ dA\ /\ 0.465^2$.

In view of the earlier discussion in this chapter, there is nothing new to be said about displacement constraints. Stress constraints are another matter. First of all, the stress constraints involve both axial load and bending. Here we have used the most simple relationship that requires that

$$\frac{axial\ load}{member\ area} + \frac{\max imum\ moment}{\sec tion\ \mathrm{mod}\, ulus} \le allowable\ stress$$

As in the case of the truss, the axial load is computed using the member length change. The maximum moment is computed by checking each end of the member and then using the relationship between moment and the displacement variables. Briefly, each member is described by three displacement variables

$$\Delta_i = \begin{bmatrix} \Delta L_i \\ \alpha_i^+ \\ \alpha_i^- \end{bmatrix}$$

from which the member forces F_i are computed as $F_i = K_i\,\Delta_i$. Here

ΔL_i – length change of member i

α_i^+ – rotation of the joint at the + end of the member minus the rigid body rotation

α_i^- – rotation of the joint at the – end of the member minus the rigid body rotation

This will be seen below as involving a linear combination of displacement variables when the code is discussed (Fig. 3.5).

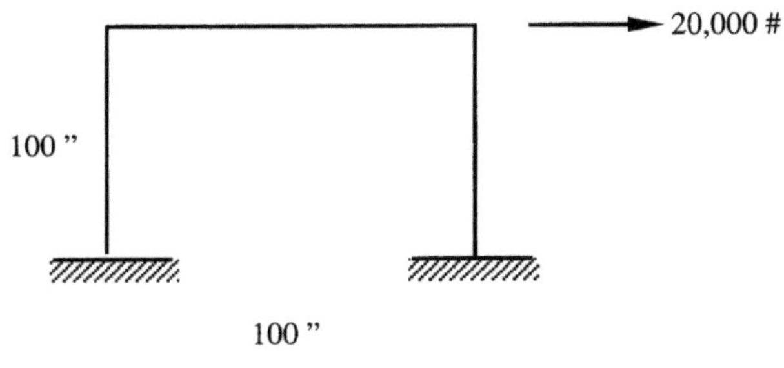

Fig. 3.5. Plane Frame

This frame example is listed on the CD as Program 12. It is built on the "frame" used above for the other programs. But there are some differences to be discussed:

- Replacing dI_i by dA_i. The term $dK\ \Delta$ in the incremental equations was a simple matter to deal with in the truss problem by scaling the equations as they go into the LP solver. Here there are two problems. First, dI_i must be replaced by dA_i as discussed above. Second, since K_i is not diagonal, it must be treated more carefully. That is done in to code by introducing a new array x2al with the property dK Δ = x2al * dA. This is used in the code by forming two new arrays prod → prodn and prod2 • prod2n:

```
      DO 22 I=1,NB3
      DO 22 J=1,NB
   22 X2AL(I,J)=0.
      DO 723 I=1,NB
      X2AL(3*I-2,I)=E*DELTA(1,3*I-2)/AL(I)
      X2AL(3*I-1,I)=E*9.184*A(I)*(4.*DELTA(1,3*I-1)+2.*
     1 DELTA(1,3*I))/AL(I)
  723 X2AL(3*I,I)=E*9.184*A(I)*(2.*DELTA(1,3*I-1)+4.*
     1 DELTA(1,3*I))/AL(I)
      DO 24 I=1,N
      DO 24 J=1,NB
      PRODN(I,J)=0.
      DO 24 K=1,NB3
   24 PRODN(I,J)=PRODN(I,J)+PROD(I,K)*X2AL(K,J)
      DO 25 I=1,NB3
      DO 25 J=1,NB
      PROD2N(I,J)=0.
      DO 25 K=1,NB3
   25 PROD2N(I,J)=PROD2N(I,J)+PROD2(I,K)*X2AL(K,J)
```

- Formation of the stress constraints. As described above, the stress constraints involve linear combinations of the displacement variables:

```
C      STRESS CONSTRAINTS
NROWS=0
DO 31 I=1,NB
DO 31 J=1,1
CALL PAPMS(FSAVE,I,SIGP,SIGM,A(I))
CONS=SIGN(1.,FSAVE(1,3*I-1))
AMOMA=FSAVE(1,3*I-1)
C1=4.
C2=2.
C11=1.
C12=0.
IF(ABS(SIGM).LE.ABS(SIGP))GO TO 344
CONS= SIGN(1.,FSAVE(1,3*I))
AMOMA=FSAVE(1,3*I)
C1=2.
C2=4.
C11=0.
C12=1.
344 CONTINUE
```

```
RAT=AMAX1(ABS(SIGP),ABS(SIGM))/SIGALL(1,I)
IF(RAT.LT..8)GO TO 31
NROWS=NROWS+1
DO 34 K=1,NB
ALP(NROWS,K)=E*PROD2N(3*I-2,K)*SIGN(1.,FSAVE(1,3*I-
 2))/AL(I)
C
1 +C1*PROD2N(3*I-1,K)*(E*AIMOD(A(I))/AL(I))/
1 SMOD(A(I))*CONS
C
1 +C2*PROD2N(3*I,K)*(E*AIMOD(A(I))/AL(I))/
1 SMOD(A(I))*CONS
34 CONTINUE
ITYPE(NROWS)=1
BU(NROWS)=SIGALL(1,I)-ABS(FSAVE(1,3*I-2))/A(I)
1-ABS(AMOMA)/SMOD(A(I))
31    CONTINUE
```

3.8 A Buckling Example

To further demonstrate the facility of the *incremental* method, this section discusses the problem of the optimal shape of a pin-ended column that was solved by Keller (1960). First, consider the case of Euler buckling of a uniform column (Fig. 3.6).

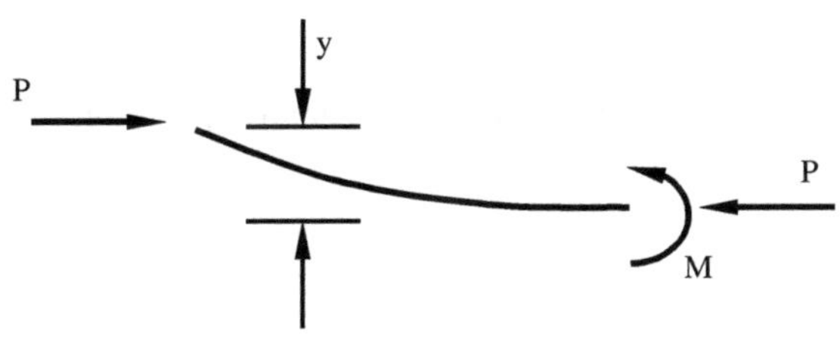

Fig. 3.6. Deformed Beam Segment

Typically, this discussion starts with the case of deformed equilibrium (Fig. 3.4) for which

$$M = P\,y$$

When the constitutive equation, $M = -EI\,y''$, is added the second-order equation

$$y'' + (P/EI)y = 0$$

results and can be solved for the buckling load $P_{CR} = \pi^2\, EI/L^2$.

Keller considers the case in which the moment of inertia I is allowed to vary. Here we propose to recompute Keller's results using the method of finite differences. For that purpose the column is divided into n+1 segments (Fig. 3.7) at which the second derivative y'' is approximated using the central difference operator

Fig. 3.7. Finite Difference Model of a Beam

$$y'' \cong \left(y^{+} + y^{-} - 2y^{0}\right)\frac{1}{H^2}$$

This leads to representing the differential equation by matrix equation

$$Ry + PSy = 0$$

Here

- y – n × 1 column matrix containing the displacements along the beam
- R – n × n tridiagonal matrix with elements $R_{i,i-1} = \alpha$, $R_{i,i} = 2\alpha$, $R_{i,i+1} = \alpha$ with $\alpha = 1/H^2$
- S – n × n diagonal matrix with elements $S_{ii} = 1/(EI)$
- P – buckling load

An analysis problem can now be stated as find P by solving the eigenvalue problem $Ry + PSy = 0$. This will be shown below to return the Euler buckling load when I is constant.

As an *optimization* problem we will now solve the problem of find I(x) to minimize the volume while keeping P constant. Again we are going to use the incremental approach. First, Keller assumes that the moment of inertia is proportional to the area squared, that is

$$A_i = I_i^{1/2} \Rightarrow dA_i = 1/2\, I_i^{-1/2}\, dI_i$$

The incremental form of the buckling equation is then

$$R\, dy + P\, dS\, y + PS\, dy = 0$$

If this equation is multiplied by y^T it follows that

$$y^T R\, dy + y^T P\, dS\, y + y^T PS\, dy = 0 \Rightarrow y^T P\, dS\, y = 0 \text{ with } dS_{ii} = -dI_i/(EI_i^2)$$

The sequential linear programming problem to be solved is then

$$\min \sum H \frac{1}{2} I_i^{-1/2} dI_i \qquad \textit{subject to} \qquad y^T PdSy = 0$$

That is done in program Prog13.for that follows:

```
C       COLUMN BUCKLING (KELLER'S PROBLEM) PROG13.FOR
        USE IMSL
        DOUBLE PRECISION D2(50,50),A(50,50),AII(50),EVAL(50)
      & ,EVEC(50,50),F(50),ALP(100,100),OBJ
      & ,XLB(50),XUB(50),BL(50),BU(50),XSOL(50),DSOL(50)
        DIMENSION ITYPE(50)
C       INITIALIZE PARAMETERS/ARRAYS
C
        E = 29.0D06
        NSPACES=20
        NODES=NSPACES-1
        ALEN=120.
        PCR=1.987E7
        H=ALEN/FLOAT(NSPACES)
        H2=H*H
        DO 3 I=1,NODES
      3 AII(I)=1000.
C
        DO 9997 ITER=1,40
        DO 1 I=1,NODES
        DO 2 J=1,NODES
        A(I,J)=0.

    2 D2(I,J)=0.
        A(I,I)=1./(E*AII(I))
        IF(I.NE.1) D2(I,I-1)=1./H2
        IF(I.NE.NODES) D2(I,I+1)=1./H2
        D2(I,I   )=-2./H2
      1 CONTINUE
C        WRITE(60,*)((I,J,A(I,J),D2(I,J),J=1,NODES),I=1,NODES)
        CALL DGVCSP(NODES,D2,50,A,50,EVAL,EVEC,50)
```

```
      WRITE(60,*)NSPACES,EVAL(NODES),(I,EVEC(I,NODES)
     & ,I=1,NODES)
C
C     SCALE FOR THE EIGENVALUE
C
      RATIOO=PCR/ABS(EVAL(NODES))
      DO 88 I=1,NODES
   88 AII(I)=RATIOO*AII(I)
C
      DO 70 I=1,NODES
      F(I)=.5/SQRT(AII(I))
      ALP(1,I)=-(EVEC(I,NODES)**2)/(AII(I)**2)
      XLB(I)=-.1*AII(I)
   70 XUB(I)=.1*AII(I)
      BL(1)=0.
      BU(1)=0.
      ITYPE(1)=3
      NROWS=1
      LDA=100
C      WRITE(60,*)' *** NROWS',NROWS
      WRITE(60,*)((ALP(I,J),J=1,NODES),I=1,NROWS)
      CALL DDLPRS(NROWS,NODES,ALP,LDA,BL,BU,F,ITYPE,
     & XLB,XUB,OBJ,XSOL,DSOL)
      WRITE(60,61)OBJ,(I,XSOL(I),AII(I),I=1,NODES)
   61 FORMAT(E20.8/ (I5,2E20.8))
      VOLUME=0.
      DO 71 I=1,NODES
      VOLUME=VOLUME+H*SQRT(AII(I))
   71 AII(I)=AII(I)+XSOL(I)
      WRITE(60,*)(I,AII(I),I=1,NODES)
      WRITE(60,*) '*** VOLUME', VOLUME,NROWS,ITER
 9997 CONTINUE
      STOP
      END
```

This program starts with a uniform column with a moment of inertia of 1000 in.^{4} and a length of 120 in. for which the buckling load is 1.987×10^7 lbs. It begins an iterative process by first computing the buckling load and then scaling the moment of inertia to return this buckling load. (The first iteration returns the buckling load for a uniform column.) Note that the smallest eigenvalue corresponds to the last one produced by the IMSL routine DGUCSP. Figure 3.8 shows the moment of inertia of the optimal column.

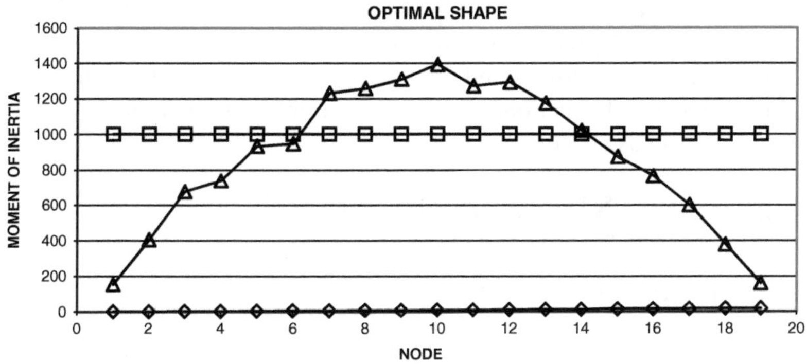

Fig. 3.8. Keller's Solution

3.9 The Incremental Equations when Shape Change is Allowed

For reference when later chapters are developed, it is convenient to point out here that the incremental equations are more complicated when the structure is allowed to change shape. In this case the matrix N must change with the changing geometry:

	Equations of structures	Incremental form
Equilibrium equations	$N^T F = P$	$N^T dF + dN^T F = 0$
Constitutive equations	$F = K \Delta$	$dF = dK \Delta + K d\Delta$
Node/member displacement	$\Delta = N \delta$	$d\Delta = N d\delta + dN \delta$

In shape optimization, the idea is to let the shape vary in addition to varying the member stiffnesses. The most simple problem of this type is the case of the statically determinate truss for which it is desired to vary the joint coordinates while minimizing the volume of material. That is, find the joint coordinates R to minimize the volume of the truss expressed as

$$minimize \quad \varphi = \sum_{members} |F_i| L_i \qquad subject\ to \qquad N^T F = P$$

In the case considered here, a statically determinate truss is considered and it is assumed that the member area is proportional to the absolute value of the member force. The only constraint is node equilibrium.

Clearly the member forces and the joint coordinates are linked and it is convenient, in the linearized equations, to use two variables, *dF* and *dR*=δ. The linearized equations in this case become

$$\textit{minimize}\ d\phi = \sum_{members}\left[L_i \operatorname{sgn} F_i \cdot dF_i + |F_i| dL_i\right]$$

$$\textit{subject to}\quad dN^T F + N^T dF = 0$$

In the above equation it only remains to express the variation of the member length dL_i in terms of the change in the node coordinates dR_i . That can be done by simple differentiation noting that

$$\partial L_i / \partial R_j = N_{i,j} \qquad or \qquad \nabla L = N$$

That is, the gradient of the member length *L* equals the generalized incidence matrix N. This follows directly from the fact that

$$L_i = \left[(x_A - x_C)^2 + (y_A - y_C)^2\right]^{1/2}$$

when member i has nodes A and C as its positive and negative ends. Then

$$\partial L_i / \partial x_A = \frac{(x_A - x_c)}{\left[(x_A - x_c)^2 + (y_A - y_c)^2\right]^{1/2}} = (n_i)_x$$

The shape optimization problem to be solved is finally, find dF_i and dR_i to

$$\textit{minimize}\ \varphi = \sum\left[(\operatorname{sgn} F_i)L_i dF_i + (N^T \mid F \mid)_i dR_i\right]$$

$$\textit{subject to}\quad K_G dR + N^T dF = 0$$

Here the geometric stiffness matrix K_G follows from the fact that in the above equation,

$$dN^T F \quad \Rightarrow \quad K_G \delta = K_G dR$$

See Appendix B and Levy and Spillers (2003) for a discussion of the geometric stiffness matrix for trusses.

There are two computer programs listed on the CD that are pertinent to this section:

1. Prog14.for runs with Prog14.dat as data
2. Prog15.for runs with Prog15.dat as data

The first of these is just a two-bar truss that reshapes into two bars at 45°. The second is a more complex example that will be discussed in some detail.

```
C      2D TRUSS GEOMETRIC OPTIMIZATION (PROG14.FOR)
C      NONLINEAR VERSION
       USE IMSL
       DIMENSION NP(100),NM(100),S(100),PSAVE(100),FSAVE(100)
      & ,ANT(50,50),P(50),CB(50,50),ANTB(50,50),BL(50),BU(50)
      & ,IRTYPE(50),XLB(50),XUB(50),XSOL(50),DSOL(50),ANTAF(50)
      & ,ALEN(50)
       DOUBLE PRECISION R(100),C(100,100),UVEC(2),C1,D1,
      1 F1,F2,FAC1,C2
       MAXC=100
C
C      INITIALIZE PARAMETERS/ARRAYS
C
       E = 30.0D06
       MAXC=100
       FAC1=1.
     5 READ(50,2,END=299) NB,NN,NS,NSTEP,NIT
       IF(NSTEP.EQ.0) NSTEP=1
       IF(NIT.EQ.0) NIT=1
     2 FORMAT (5I5)
       WRITE(60,1)NB,NN,NS,NSTEP,NIT
     1 FORMAT(I5,'  NO. MEMBERS'/I5,'  NO. NODES'/I5,
      & '  NO. SUPPORTS'
      1 /I5,'  NO. LOAD STEPS'/I5,'  NO.ITERATIONS'//)
       NNS=NN-NS
       N=2*NNS
       NMOV=6
       READ(50,156)(R(2*K-1),R(2*K),PSAVE(2*K-1),
      1 PSAVE(2*K),K=1,NN)
       WRITE(60,157)(K,R(2*K-1),R(2*K),PSAVE(2*K-1),
      1 PSAVE(2*K),K=1,NN)
157    FORMAT (1H1,25X,11HCOORDINATES,40X,5HLOADS//
      114X,1HX,19X,1HY,18X,2HPX,18X,2HPY,//(I4,4D20.8))
156    FORMAT (8X,4F11.6)
       WRITE(60,159)
159    FORMAT (1H1,3X,6HMEMBER,5X,5H+ END,
      & 5X,5H- END,16X,4HAREA,11X,
      1 9HPRESTRESS//)
       READ(50,151)(NP(L),NM(L),S(L),FSAVE(L),L=1,NB)
       WRITE(60,160)(L,NP(L),NM(L),S(L),FSAVE(L),L=1,NB)
151    FORMAT (2I5,8X,2E10.6)
160    FORMAT (3I10,2E20.8)
C
C      START LOADSTEPS AND ITERATIONS
       NIT=20
       DO 997 ITER=1,NIT
       WRITE(60,897) ITER,LDSTP
   897 FORMAT(///' ****ITERATION NUMBER',I4/
```

```
     1          '       LOAD STEP       ',I4)
C
C      SET UP SYSTEM MATRIX
C
       DO 904 I=1,N
       P(I)=PSAVE(I)
       DO 904 J=1,N
       ANT(I,J)=0.
   904 C(I,J)=0.
       DO 999 L=1,NB
       S(L)=0.
       K = 2*NP(L)
       M = 2*NM(L)
       CALL UNITV(K,M,C1,UVEC,R)
       ALEN(L)=C1
       IF(K.GT.N) GO TO 888
       ANT(K-1,L)=UVEC(1)
       ANT(K   ,L)=UVEC(2)
   888 IF(M.GT.N) GO TO 887
       ANT(M-1,L)=-UVEC(1)
       ANT(M   ,L)=-UVEC(2)
   887 CALL INSERT(C,K,M,UVEC,MAXC,N,E,S(L),C1,FSAVE(L))
  999  CONTINUE
       DO 3 I=1,N
       DO 3 J=1,N
       ANTB(I,J)=ANT(I,J)
     3 CB(I,J)=C(I,J)
       DO 33 I=1,N
       DO 33 J=1,NB
    33 CB(I,J+NMOV)=ANT(I,J)
       IPATH=1
       LDA=50
       NVAR=NMOV+NB
       CALL LSLRG(N,ANTB,LDA,P,IPATH,FSAVE)
       WRITE(60,*)(I,FSAVE(I),I=1,NB)
       IF(ITER.EQ.1) GO TO 997
       DO 11 I=1,N
       ANTAF(I)=0.
       BU(I)=0.
       BL(I)=0.
       IRTYPE(I)=3
       DO 11 J=1,NB
    11 ANTAF(I)=ANTAF(I)+ANT(I,J)*ABS(FSAVE(J))
       NMOV1=NMOV+1
       DO 111 I=NMOV1,NVAR

  111 ANTAF(I)=ALEN(I-NMOV)*SIGN(1.,FSAVE(I-NMOV))
       DO 4 I=1,NMOV
       XUB(I)= .1*R(I)
     4 XLB(I)=-.1*R(I)
       NMOV1=NMOV+1
       DO 44 I=NMOV1,NVAR
       XUB(I)= 1.*ABS(FSAVE(I-NMOV))
    44 XLB(I)=-1.*ABS(FSAVE(I-NMOV))
       WRITE(60,*)((I,J,CB(I,J),J=1,NVAR),I=1,NMOV)
       WRITE(60,*)(I,ANTAF(I),I=1,NVAR)
       CALL DLPRS(N,NVAR,CB,LDA,BL,BU,ANTAF,IRTYPE,XLB,
      & XUB,OBJ,XSOL,DSOL)
```

```
C
C
  998 CONTINUE
C
C       UPDATE COORDINATES
        CON=-1.
        DO 444 I=1,NMOV
        CON=-1.*CON
        WRITE(60,*) XSOL(I)
  444 IF(CON.LT.0.) R(I)=R(I)+XSOL(I)
        R(6)=R(2)
        VOL=0.
        NB1=NB-1
        DO 944 I=1,NB1
  944 VOL=VOL+ALEN(I)*ABS(FSAVE(I))
        WRITE(60,*) 'VOLUME', VOL
        WRITE(60,158)(K,R(2*K-1),R(2*K),K=1,NNS)
158     FORMAT (1H1,11X,19HUPDATED COORDINATES//
       114X,1HX,19X,1HY,//(I4,2D20.8))
  997 CONTINUE
  299 STOP
        END
```

After some input, the program proceeds in a straightforward manner:

1. The matrix N^T is computed as ANT.
2. The linear programming matrices CB, ANTAF, etc., are computed.
3. The coordinates are updated (Fig. 3.9).

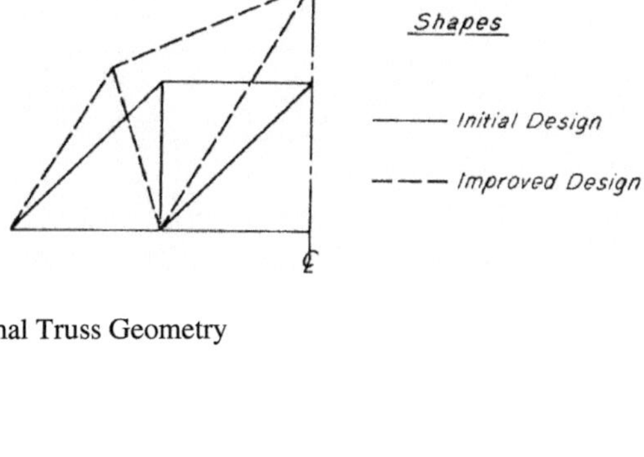

Fig. 3.9. Optimal Truss Geometry

3.10 A Beam Example

Figure 3.10 shows the optimal design for a beam under a uniform load using the incremental approach and linear programming. This is the problem of finding the optimal moment diagram to

$$minimize \quad \int | M | \, dx \qquad subject \ to \qquad M'' = -w$$

Here it is assumed that the cost of the beam section is proportional to the absolute value of the moment and that equilibrium must be satisfied. This is of course plastic design.

A finite difference solution using Matlab is given. The beam centerline is described by n nodes and the central difference operator is used for the second derivative. (See the column buckling problem above.) The incremental version of the problem is then to find dM

$$minimize \quad \int | M + dM | \, dx \qquad subject \ to \quad (M + dM)'' = -w$$

Looking at the code it can be seen that the identity

$|x + dx| = |x| + dx \, sgn(x) \quad dx << 1$

is used.

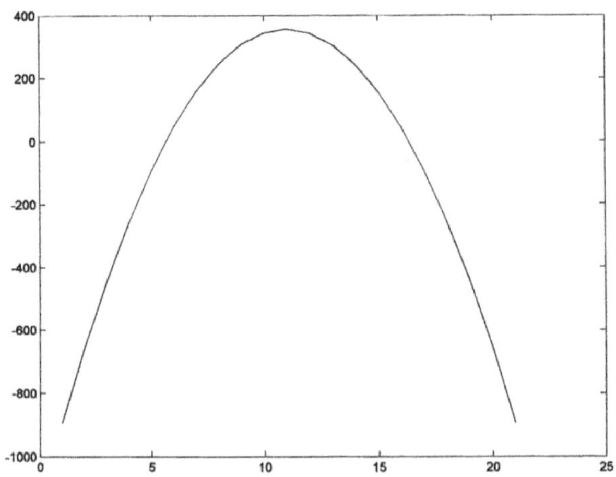

Fig. 3.10. Optimal Beam Design

Program 16

```
%Optimum Fixed Beam - Prog16.m
clear all
alen=100.;
nit=10;
nodes=21; %must be an odd number
nseg=nodes-1;
dx=alen/(nseg)
small=1.;
inodes1=nodes-2;
amom=zeros(nodes,1);
lb=-6.*ones(nodes,1);
ub=+6.*ones(nodes,1);
xx=0;
%Use initial solution for a fixed beam
for i=1:nodes
    amom(i)=(-alen^2/12.+xx*alen/2.-xx*xx/2.);
    xx=xx+dx;
end
%Finite difference solution
for iter=1:nit
    iter
    fac=fac*1.;
    Aeq=zeros(inodes1,nodes);
    beq=zeros(inodes1,1);
    for i=1:inodes1
        beq(i)=-1.;
        Aeq(i,i+1)=-2./(dx*dx);
        Aeq(i,i+2)=1./(dx*dx);
        Aeq(i,i)=1./(dx*dx);
        beq(i)=beq(i)+amom(i+1)*2./(dx*dx);
        beq(i)=beq(i)-amom(i+2)*1./(dx*dx);
        beq(i)=beq(i)-amom(i)*1./(dx*dx);
        end;
        %
    for i=1:nodes
        f(i)=sign(amom(i))*dx;
    end
        f(1)=f(1)/2.
        f(nodes)=f(nodes)/2.

[x,fval,exitflag,output,lambda]=linprog(f,[],[],Aeq,beq,lb,ub)
    %Increment the moments
    for j=1:nodes
        amom(j)=amom(j)+x(j);
    end
end
%Plot moment diagram
    x=1:1:nodes;
    amom
    plot(x,amom(x))
    set(findobj('Type','line'),'Color','k')
    pause
```

3.11 A Plate Bending Problem

In the spirit of the beam problem just solved, the final example of this chapter is a plate bending problem of finding M_x , M_y , and M_{xy} to

$$\text{minimize} \quad \int_S [M_x^{\,2} - M_x M_y + M_y^{\,2} + 3M_{xy}^2]^{1/2}$$

$$\text{subject to} \quad M_{x,xx} + M_{y,yy} - 2M_{xy,xy} = -q$$

This problem attempts to minimize the material volume written using the Mises yield condition while satisfying the equations of equilibrium.

This problem is solved using Matlab and listed with the other programs as plate.m. A finite difference solution is again used dividing the plate into a number of nodes. The square plate with a uniform load is considered; symmetry is used reducing the solution to ¼ of the plate. The finite difference solution again uses the second derivative as do some earlier problems. It also uses the mixed derivative indicated in Fig. 3.14.

Program Prog17.m is longer than the beam problem described above and its details are left to the reader. Figures 3.11–3.13 show the optimal moment diagrams.

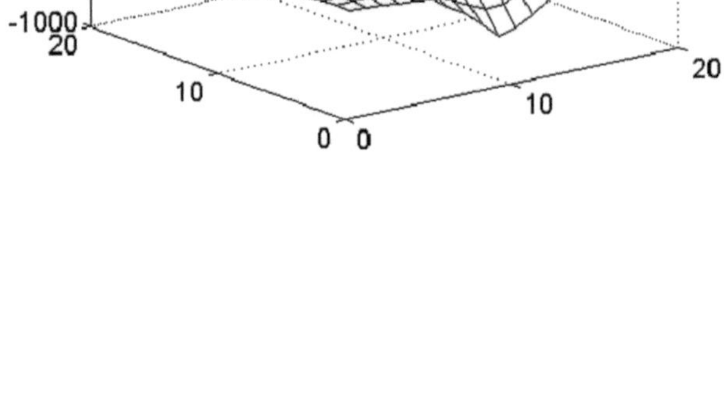

Figure 3.11
M_x

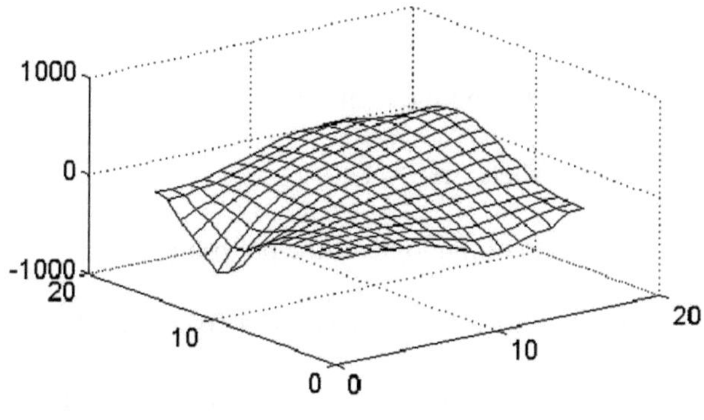

Figure 3.12
M_y

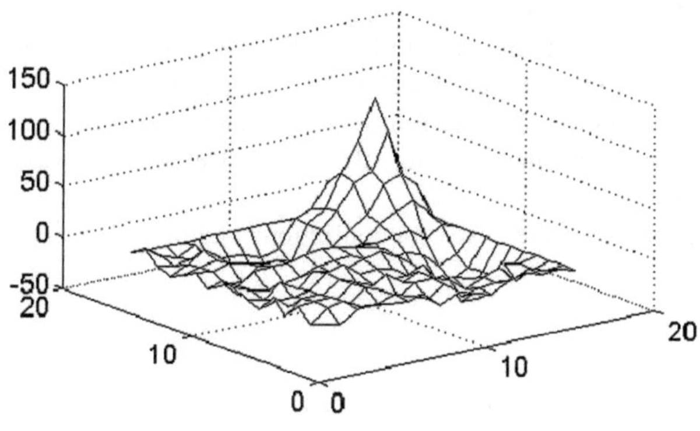

Figure 3.13
M_{xy}

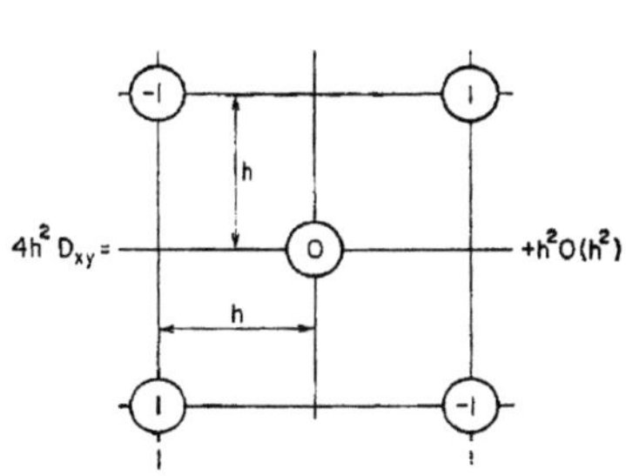

Fig. 3.14. Finite Difference Representation of the Mixed Derivative

3.12 Problems

1. Rewrite the code of Program 9 using the identity

   $$| x + dx | = | x | + \text{sgn}(x)\, dx \quad \text{when} \quad dx << 1.$$

   This will allow a simplification of the logic of the program.

2. Redo the buckling example above for the case of a fixed-ended column.
3. In the 25-bar truss, try using code-based stress constraints rather than constant allowable stresses.
4. Rewrite the plate problem using 1/8th symmetry rather than ¼ symmetry.
5. Rewrite the optimal truss problem of shape optimization using code-based constraints that include the effect of buckling of members under compression. (We will return to that problem in a later chapter.)
6. Redo the buckling example using the fourth-order beam equations.
7. Redo one of the truss problems using allowable stresses from the AISC specifications.
8. Redo the beam problem for the case of a continuous beam.

4 Optimality Criteria Methods

This chapter discusses the so-called optimality criteria methods of structural optimization. They are generally regarded to be heuristic when compared to the more formal methods of mathematical programming. Examples of both discrete and continuous systems are considered. These include trusses, frames, and plates.

4.1 Introduction

Clearly, the response of an elastic structure is controlled by its member stiffnesses. Historically, the engineer looks at his/her design and modifies the member stiffnesses when some improved performance is thought to be needed. Then a reanalysis is carried out and the response is again studied. This is the manual analysis/redesign process. Optimality criteria methods mirror this process. (How should the member stiffnesses be adjusted given some desired change in response?) And while the methods of mathematical programming are formal in the sense of mathematics, optimality criteria methods are considered to be *heuristic*.

Historically, there has been a tension in structural optimization between what have been called *optimality criteria methods* and the classical methods of mathematical programming: The classical group has complained that the optimality criteria workers lack theory and the optimality criteria group has argued that the classical workers have been unable to solve real (large) problems. As more and more optimization tools become available, it would appear that this latter comment would at some point cease to pertain. In any case, optimality criteria methods play an important role in structural optimization. This chapter attempts an introduction to these optimality criteria methods.

With classical optimization, problems are described in terms of objective functions and constraints and then solved using some mathematical programming algorithm. With optimality criteria methods, a structure is analyzed and then redesigned on the basis of some *resizing rule*. (Optimality criteria methods are typically *secant methods* while mathematical programming methods are typically incremental methods.) Since these rules can be somewhat heuristic, the designer may not be quite sure what he/she is getting. On the other hand, the designer can always monitor the objective function to verify that it is behaving properly.

Optimality criteria methods have been championed by Venkayya (1971, 1973) and his co-workers at Wright-Patterson AFB in publications going back to the 1960s. A good overview of optimality criteria methods can be found in Kirsch (1981). (See also Spillers, 1975).

W.R. Spillers, K.M. MacBain, *Structural Optimization*,
DOI 10.1007/978-0-387-95865-1_4,

4.2 The Most Simple Optimality Criteria Problem

All optimality criteria methods are based on the common analysis/redesign procedure used historically by engineers. For the truss problem this involves iterating

$$A_i^{(n+1)} = | F_i^{(n)} | / \sigma_a$$

This implies that at some iteration in the design process the engineer, after n iterations, is still not happy with his/her design and selects a new bar area $A_i^{(n+1)}$ in an attempt to satisfy allowable stress criteria in which the bar force F_i is fully stressed with allowable stress σ_a. This step is then followed by a reanalysis. (Note that the allowable stress is taken to be constant here and does not reflect the buckling penalty, for example, of the AISC code for bars in compression.) It will now be shown that the process of doing so behaves monotonically when the allowable stress is constant.

Consider the following *primal* problem: Find F to

$$\text{minimize } |F| \cdot \Delta_a \qquad \text{subject to} \qquad N^T F = P \;\; \text{(node equilibrium)}$$

Where F is the member force matrix, P is the node load matrix, and Δ_a the matrix of member allowable length change, that is, the length change of the bar when it is fully stressed. (F_i is the force in truss bar i, $(\Delta_a)_i = \sigma_i L_i / E$, L_i is the length of member i, and E is Young's modulus.) A hybrid *dot product* notation has been used here so that $|F| \cdot \Delta_a = \Sigma |F_i| (\Delta_a)_i$. This sum is proportional to the volume of material when $|F_i|$ is proportional to the member area and σ_a, the allowable stress, is constant. This *primal* problem then looks for minimum weight and tries to satisfy the equations of node equilibrium. There are several possible interpretations of this *linear programming* problem. With the most simple, plastic design, the member area is proportional to $|F_i|$ so that the objective function is proportional to the volume of material in the structure. As an elastic design problem the member area is again argued to be proportional to $|F_i|$ but the neglect of any discussion of displacements requires the idea of *embedding*. The units of the objective function have been selected to be those of energy. This will be seen to be convenient in the work that follows.

The *dual* problem (see the exercises at the end of this chapter) then has the form, find the joint displacement matrix δ to

$$\text{maximize } P^T \delta \quad \text{subject to} \quad |\Delta| = |N\delta| \leq \Delta_a$$

That is, find the joint displacement matrix δ to maximize the work done by the joint loads P while keeping the member displacement matrix Δ (the member length change in the case of the truss) less than the allowable.

Note that the important *continued inequality*

$$|F| \cdot \Delta_a \geq |F| \cdot |\Delta| \geq F \cdot \Delta = P \cdot \delta$$

holds. (The above inequality uses the equation of virtual work, $F \cdot \Delta = P \cdot \delta$.)
This primal problem can be solved iterating

$$K_i^{(N+1)} = | F_i^{(N)} | /(\Delta_a)_i \qquad \textit{with} \qquad F^{(N)} = K^{(N)} N \left(N^T K^{(N)} N\right)^{-1} P$$

Put simply, this algorithm computes the member forces using the equations of the node method while trying to satisfy allowable stress criteria at each iteration. It has broad utility which will be discussed below. For example, it can be used to select an optimal subset from an initial structure containing too many nodes and bars.

4.3 Monotone Behavior

The algorithm just described will now be shown to behave monotonically. Let

$$\varphi^{(N)} = | F^{(N)} | \cdot \Delta_a = | \Delta^{(N)} | \cdot K^{(N)} \cdot \Delta_a = \Delta_a \cdot K^{(N+1)} \cdot \Delta_a$$

using the constitutive equation and the resizing rule. It is useful now to develop the following *virtual work* lemma:

Lemma: $\Delta^{(N)} \cdot K^{(N)} \cdot \Delta^{(N)} = \Delta^{(N)} \cdot K^{(N-1)} \cdot \Delta^{(N-1)}$

Proof: Since both $F^{(N)}$ and $F^{(N-1)}$ satisfy node equilibrium, it follows that

$$N^T F^{(N)} = N^T F^{(N-1)} = P \qquad \Leftrightarrow \qquad \delta^{(N)T} N^T F^{(N)} = \delta^{(N)T} N^T F^{(N-1)}$$

Using the constitutive equation $F = K\Delta$ and the member/node displacement equation $\Delta = N\,\delta$ the lemma follows directly.

Theorem. (Monotone Behavior) $\varphi^{(N)} \leq \varphi^{(N-1)}$

Proof:

$$\begin{aligned}
(\varphi^{(N)})^2 &= (| \Delta^{(N)} | \cdot K^{(N)} \cdot \Delta_a)^2 \\
&\leq (\Delta_a \cdot K^{(N)} \cdot \Delta_a)\,(| \Delta^{(N)} | \cdot K^{(N)} \cdot | \Delta^{(N)} |) && \text{(using Schwarz's inequality)} \\
&= \varphi^{(N-1)}\,(\Delta^{(N)} \cdot K^{(N)} \cdot \Delta^{(N)}) \\
&= \varphi^{(N-1)}\,(\Delta^{(N)} \cdot K^{(N-1)} \cdot \Delta^{(N-1)}) && \text{(using the lemma)} \\
&\leq \varphi^{(N-1)}\,(|\Delta^{(N)}| \cdot K^{(N-1)} \cdot |\Delta^{(N-1)}|) \\
&= \varphi^{(N-1)}\,(|\Delta^{(N)}| \cdot K^{(N)} \cdot \Delta_a) && \text{using resizing algorithm}
\end{aligned}$$

$$= \varphi^{(N-1)} \varphi^{(N)}$$

or

$$\varphi^{(N-1)} \leq \varphi^{(N)}$$

4.4 An Application

One application of this method is selecting a design for the engineer (see Figs. 4.1 and 4.2). In this case, a grid is laid out and every node is connected to every other node. As the algorithm iterates it selects a statically determinate substructure which is an optimal design. (This can be argued, for example, using the fact that a linear programming solution is *basic*.) Presumably, there are cases in which the design selected contains surprises for the engineer. Program 18 on the CD indicates how simple it is to do this kind of design.

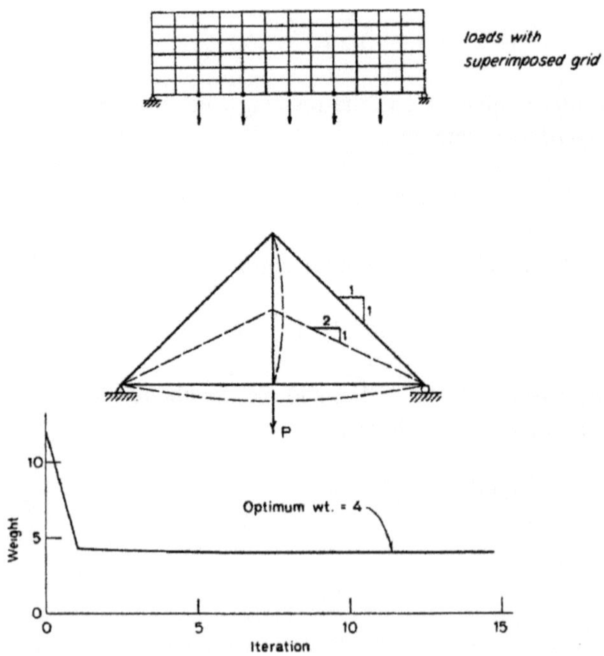

Fig. 4.1. Using a Grid to Create Designs

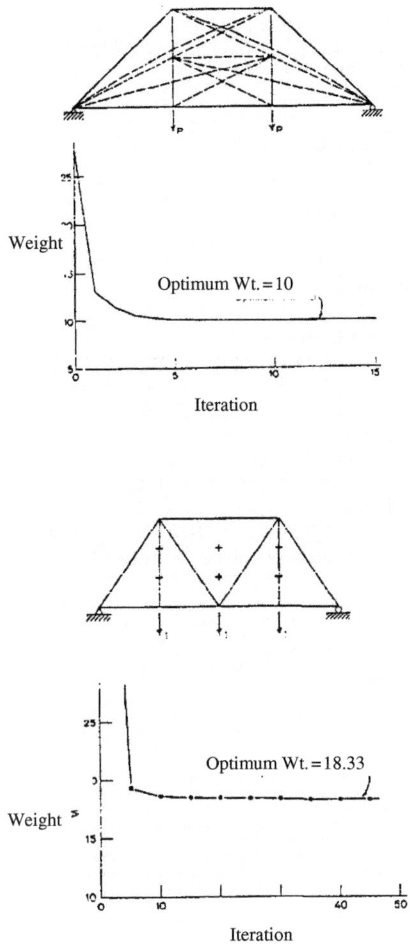

Fig. 4.2. Selecting an Optimal Substructure

The following partial (some basic subroutines are omitted, see Appendix B) listing of Program 18 indicates how simple iterative design can be: (1) the program must be made to iterate which means that the joint loads must be saved and (2) the bar areas are modified at each iteration to satisfy allowable stress criteria. Note that if you are not careful, some bar areas going to zero may ill condition the system matrix.

```
C      PROGRAM 18. TR2D.FOR
C      PLANE TRUSS MODIFIED FOR ITERATIVE DESIGN
       DIMENSION NP(100),NM(100),S(100)
       DOUBLE PRECISION R(100),P(100),C(100,100),UVEC(2)
      1,C1,D1,F1,F2,FAC,PSAVE(100)
       MAXC=100
C
C      INITIALIZE PARAMETERS/ARRAYS
C
       E = 29.0D06

       SIGA=1.
       SMALL=1.E-10
100    READ(50,150)NB,NN,NS,nbsup
150    FORMAT (4I5)
       WRITE(60,1)NB,NN,NS,nbsup
     1 FORMAT(I5,'  NO. MEMBERS'/I5,'  NO. NODES'/I5,
      1'  NO.SUPPORTS'/i5,'  NO.Support BARS'//)
       READ(50,156)(R(2*K-1),R(2*K),PSAVE(2*K-1),
      1 PSAVE(2*K),K=1,NN)
   156 FORMAT (8X,4F11.6)
       WRITE(60,157)(K,R(2*K-1),R(2*K),PSAVE(2*K-1),
      1 PSAVE(2*K),K=1,NN)
157    FORMAT (1H1,16X,11HCOORDINATES,28X,5HLOADS/
      1 14X,1HX,17X,1HY,16X,2HPX,16X,2HPY//(I4,4D18.8))
       NNN = NN - NS
       N=2*NNN
       nb1=nb-nbsup
C
C      SET UP SYSTEM MATRIX
C
       NIT=20
       DO 9981 ITER=1,NIT
       VOLUME=0.
       DO 30 I = 1,N
       P(I)=PSAVE(I)
       DO 30 J = 1,N
    30 C(I,J) = 0.
       WRITE(60,159)
       DO 999 L=1,NB
       IF(ITER.EQ.1)
      1   READ(50,151)NP(L),NM(L),S(L)
         WRITE(60,160)L,NP(L),NM(L),S(L)
151      FORMAT (2I5,8X,E10.6)
160      FORMAT (3I10,E20.8)
         K = 2*NP(L)
         M = 2*NM(L)
         CALL UNITV(K,M,C1,UVEC,R)
         CALL INSERT(C,K,M,UVEC,MAXC,N,E,S(L),C1)
 999   CONTINUE
C
C      SOLVE FOR JOINT DISPLACEMENTS
C
       M = N - 1
       DO 17 I = 1,M
         L = I + 1
         DO 17 J = L,N
           IF(C(J,I)) 19,17,19
   19        DO 18 K = L,N
```

```
18          C(J,K) = C(J,K) - C(I,K)*C(J,I)/C(I,I)
        P(J) = P(J) - P(I)*C(J,I)/C(I,I)
   17 CONTINUE
      P(N) = P(N)/C(N,N)
      DO 20 I = 1,M
        K = N - I
        L = K + 1
        DO 21 J = L,N
   21     P(K) = P(K) - P(J)*C(K,J)
        P(K) = P(K)/C(K,K)
   20 CONTINUE
      WRITE(60,161)(I,P(2*I-1),P(2*I),I=1,NN)
  161 FORMAT (1H1,13HDISPLACEMENTS/20X,1HX,19X,1HY//
     1//(I10,2D20.8))
      WRITE(60,162)
  162 FORMAT(1H1,3X,6HMEMBER,9X,2HDL,17X,5HFORCE,
     1 14X,6HSTRESS//)
C
C     COMPUTE MEMBER FORCES AND DISPLACEMENTS
C
      DO 998 I=1,NB
        K = 2*NP(I)
        M = 2*NM(I)
        CALL UNITV(K,M,C1,UVEC,R)
        K1=K
        D1=0.
        FAC=1.
        DO 997 J=1,2
          IF(K1.GT.N) GO TO 996
          D1=D1+FAC*(P(K1-1)*UVEC(1)+P(K1)*UVEC(2))
996       FAC=-1.
          K1=M
997     CONTINUE
        F1=D1*E*S(I)/C1
        F2=F1/S(I)
        if(i.gt.nb1)go to 9982
        VOLUME=VOLUME+S(I)*C1
        S(I)=ABS(F1)/SIGA
        IF(S(I).EQ.0.) S(I)=SMALL
9982 WRITE(60,1000) I,D1,F1,F2
 998 CONTINUE
9981 WRITE(60,61) ITER, VOLUME
 61  FORMAT('**** ITER = ',I5,'     VOLUME = ',E20.8)
     STOP
1000 FORMAT (I10,3D20.8)
 159 FORMAT (1H1,3X,6HMEMBER,5X,5H+ END,5X,5H- END,6X,4HAREA//)
     END
```

4.5 Sandwich Beams

It will be simply noted here that the sandwich beam (Fig. 4.3) is quite similar to the truss problem when it comes to the analysis of iterative design and monotone behavior (Spillers and Farrell, 1969). It is something like a continuous version of the truss in this respect.

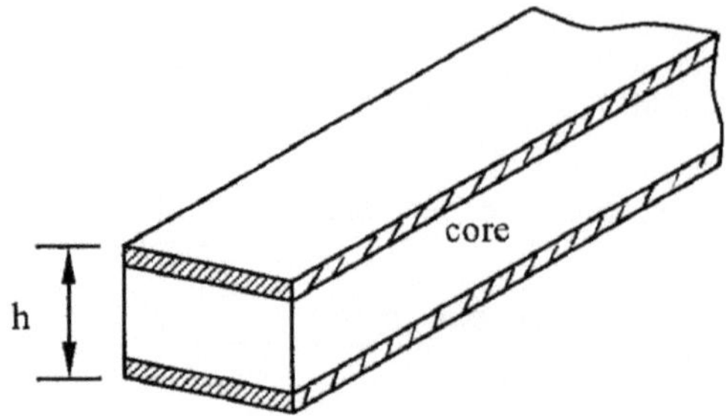

Fig. 4.3. A Sandwich Beam

4.6 A Generalization of the Truss Problem

In order to extend the discussion of the truss problem to other, more complex classes of structures, consider the problem

minimize t(F) subject to $N^T F = P$ (Node equilibrium)

with t(F) convex and homogeneous of degree 1. (This problem implies that a member design can be carried out when the member forces are given.) Applying the Lagrange multiplier method to this system gives

$$L = t(F) + \delta^T \left(P - N^T F\right)$$

$$\frac{\partial L}{\partial \delta_i} = 0 \quad \Rightarrow \quad N^T F = P$$

$$\frac{\partial L}{\partial F_i} = 0 \quad \Rightarrow \quad \nabla t = N\delta$$

An algorithm will now be developed based upon this *nonlinear* system. Note first that the only nonlinear term here is ∇t. It would be natural to linearize ∇t and then solve the system using Newton's method. But since t is homogeneous of degree 1, the Hessian matrix of t is singular. In order to avoid this problem the nonlinear system is modified to

$$N^T F = P \qquad \text{and} \qquad 2t\, \nabla t = 2t\, N\, \delta$$

by multiplying the second equation by 2t.

Let $\varphi = t^2$. Then $2t\nabla t = \nabla\varphi$ which is homogeneous of degree 1. The Hessian matrix H (matrix of second derivatives) of φ is then positive definite since t is convex. It follows that

$$\nabla\varphi(F) \approx \nabla\varphi(F_0) + H(F_0)\,(F - F_0) = H(F_0)\, F$$

The equation $\nabla t = N\,\delta \;\rightarrow\; \nabla\varphi = 2t\, N\,\delta \;\rightarrow\; HF = 2t\, N\,\delta$. Since the equations of the node method can be combined to give $F = KN\delta$, an approximate member stiffness has been derived to be

$$K = 2t\, H^{-1}$$

This is an important result (the member stiffness K is derived from the objective function t). It will be applied to several different classes of structures below. Here it will only be noted that for a single truss bar

$$(t_I)^2 = F_i^2\, \Delta_a^2 \quad \Rightarrow \quad H_{ii} = \partial^2(t_I)/\, \partial F_I^2 = 2\,\Delta_a^2 \quad \Rightarrow \quad K_{ii} = |F_i|/\Delta_a$$

This is the algorithm that was discussed at the beginning of this chapter.

For later reference, the dual problem will be discussed now briefly. It is more difficult and emphasizes the work of Lorch (1951). Lorch shows that as F ranges over all values other than zero, $\nabla t(F)$ describes a convex surface s, the adjoint surface. The interior of this surface will be designated by the set s^*. The dual problem can then be formulated as find δ to

$$\text{maximize} \quad P^T\delta \quad \text{subject to} \quad N\delta \in s^*$$

This problem has the form of maximizing the work done by the joint forces while restricting the member length displacement variable. Duality is justified by the continued inequality

$$t(F) = F\cdot\nabla t(F) \geq F\cdot\nabla t\left(\Delta^*\right) \geq F\cdot N\delta = P\cdot\delta$$

This inequality, first of all, uses the fact that t is convex and homogeneous of degree 1. It follows that given any F and Δ^*,

$$F \cdot \nabla t(F) \geq F \cdot \nabla t(\Delta^*)$$

Furthermore, if Δ^* is selected so that $\nabla t(\Delta^*)$ is parallel to $\Delta = N\delta$ and if $\Delta \in s^*$, (that is, δ is feasible) then $\|\Delta\| \leq \|\nabla t(\Delta^*)\|$. Optimality for the dual then implies that $\nabla t(F) = \nabla t(\Delta) = N\delta$ since the inequality must become an equality.

4.7 Plastic Design of Frames

This section returns to the problem of the plastic design of frames. It will be seen (Fig. 4.4) that the objective function of the problem here returns the same value previously achieved. But the frame design (the moment diagram) achieved is quite different from the earlier design. Clearly, the solution of this problem is not unique.

The frame design problem is again to find F to

$$\textit{minimize} \quad \varphi = \sum L_i \max\{|m_i^+|, |m_i^-|\} \quad \textit{subject to} \quad N^T F = P$$

with m_i^+ and m_i^- the moments at the ends of member i. There are two additional issues to be dealt with here:

1. The objective function does not include the thrust. That will be handled below by using a fictitious, large member area during the iterative process. Doing so comes from the fact that since the objective function is not a function of the member thrust, there are associated displacement terms such that $(N\delta)_i = 0$.
2. The max function is not smooth which means that its Hessian matrix is discontinuous. This can be dealt with by using the fact that for a large, even integer M,

$$\max\{|x|, |y|\} \sim \left[x^M + y^M\right]^{1/M}$$

Pursuing the details of this problem a little further, let the member force matrix F be partitioned as

$$F = \begin{bmatrix} F_1 \\ F_2 \\ \vdots \end{bmatrix} \qquad \textit{with} \qquad F_i = \begin{bmatrix} thrust_i \\ m_i^+ \\ m_i^- \end{bmatrix}$$

Then

$$t_i = \left\{ |m_i^+|^M + |m_i^-|^M \right\}^{1/M}$$

$$\frac{\partial t}{\partial m_i^+} = \left\{ |m_i^+|^M + |m_i^-|^M \right\}^{1/M-1} \left(m_i^+\right)^{M-1}$$

$$\frac{\partial t}{\partial m_i^-} = \left\{ |m_i^+|^M + |m_i^-|^M \right\}^{1/M-1} \left(m_i^-\right)^{M-1}$$

$$\frac{\partial^2 t_i}{\partial m_i^{+2}} = (M-1)\left(m_i^+\right)^{M-2} \left\{ |m_i^+|^M + |m_i^-|^M \right\}^{1/M-2} \left(m_i^-\right)^M$$

$$\frac{\partial^2 t_i}{\partial m_i^+ \partial m_i^-} = \left(m_i^+ m_i^-\right)^{M-1} (1-M) \left\{ |m_i^+|^M + |m_i^-|^M \right\}^{1/M-2}$$

Given these terms, it is a simple matter to construct the required stiffness matrix for the frame program. This stiffness matrix is used in Program 19

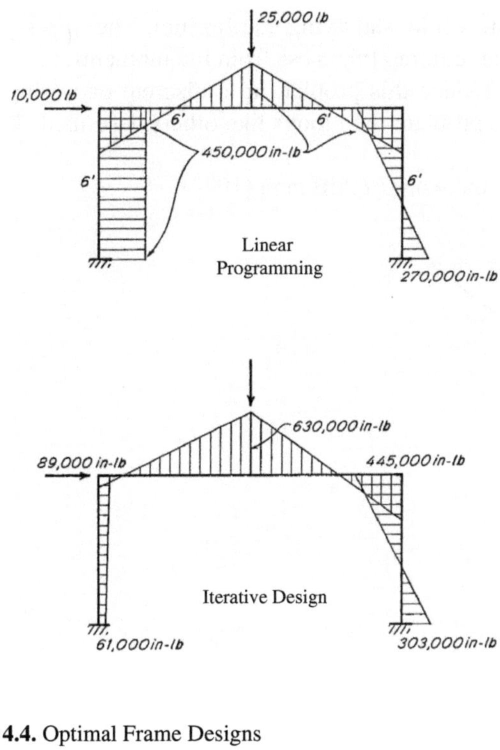

Fig. 4.4. Optimal Frame Designs

4.8 Sandwich Plate Design

A sandwich plate design problem can be formulated as follows (Hodge 1959):

$$\textit{minimize} \quad \int_S T(M)ds \qquad \textit{subject to} \qquad \Im M = q \text{ (equilibrium)}$$

where

$$T(M) = \left(m_x^2 - m_x m_y + m_y^2 + 3m_{xy}^2\right)^{1/2}$$

$$M = \begin{bmatrix} m_x \\ m_y \\ m_{xy} \end{bmatrix} \qquad \Im = (\partial^2/\partial x^2, \partial^2/\partial y^2, \partial^2/\partial x\partial y)$$

Just as in the case of the sandwich beam described above, the intention here is to minimize the volume of the flanges as represented by the thickness T(M) integrated over the surface of the plate while satisfying equilibrium. The Mises yield condition is used here to obtain the required thickness from the moments.

It is first of all convenient to reduce this problem to a discrete one using the methods of finite differences. The problem then looks like others described above. The solution follows directly.

Figures 4.5 and 4.6 come from the work of AlBanna (1972).

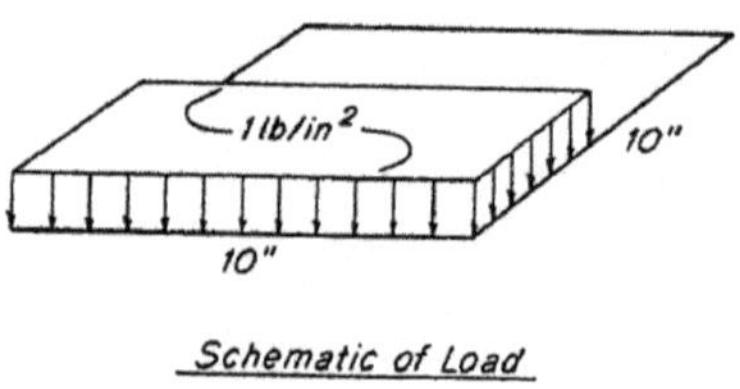

Fig. 4.5. A Sandwich Plate

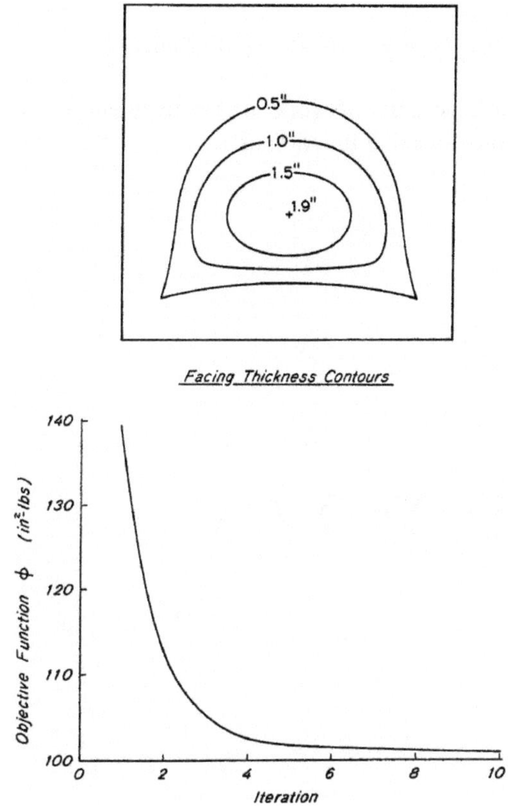

Fig. 4.6. Optimal Design of a Sandwich Plate

4.9 Truss Design

This section generalizes the earlier truss model to the case in which the allowable stress is not constant. Let σ_i be the allowable stress in member i. Safety requires that

$$A_i \geq | F_i | / \sigma_i$$

In this case the optimization problem becomes

minimize $\sum |F_i| L_i / \sigma_i$ subject to $N^T F = P$ (Node equilibrium)

In order that the objective function have units of work, it is convenient to introduce an average stress σ and write the objective function as

$$\frac{\sigma}{E} \sum |F_i| L_i \frac{\sigma}{\sigma_i}$$

or simply as

$$\sum |F_i| \overline{\Delta}_i$$

where $\overline{\Delta}_i = L_i \left(\frac{\sigma}{E}\right)\left(\frac{\sigma}{\sigma_i}\right)$. Then $t = \sum t_i = \sum |F_i| \overline{\Delta}_i$

In this case,

$$t_i = |F_i| \overline{\Delta}_i$$

$$\nabla t_i = \overline{\Delta}_i \operatorname{sgn} F_i$$

$$2t_i \nabla t_i = 2F_i \overline{\Delta}_i^2 = \nabla \phi_i$$

$$H_{ii} = 2\overline{\Delta}_i^2$$

$$K_{ii}^{(N+1)} = \frac{2\sqrt{\phi_i^{(N)}}}{H_{ii}^{(N)}} = \frac{|F_i^{(N)}|}{\overline{\Delta}_i}$$

For the case of constant allowable stresses in which $\sigma = \sigma_i$, for all i, and using the notation from elastic analysis in which

$$K_{ii} = A_i E / L_i$$

it follows that

$$K_{ii}^{(N+1)} = |F^{(N)}| / \overline{\Delta}_i = A^{(N+1)} E / L_i$$

and thus

$$A_i^{(N+1)} = \frac{L_i}{E} \frac{|F_i^{(N)}|}{\overline{\Delta}_i} = \frac{|F_i^{(N)}|}{\sigma}$$

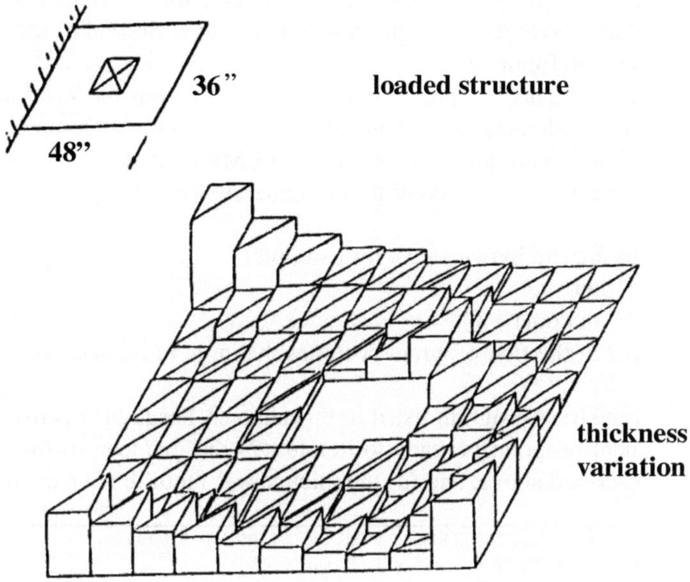

Fig. 4.7. An Optimal Plane Stress Design

4.10 A Plane Stress Problem

This section will only comment briefly on a simple finite element application (see Fig. 4.7). If you take Zienkiewicz's triangular, constant stress element and change the element stiffness matrix D as discussed above using the Hessian matrix

$$H = \begin{bmatrix} 2 & -1 & 0 \\ -1 & 2 & 0 \\ 0 & 0 & 6 \end{bmatrix} \frac{1}{\sigma^2}$$

the plate design problem takes on the form of the original truss problem discussed in this chapter. Here the thickness t is simply

$$t^2 = (F_x^2 + F_y^2 - F_x F_y + 3 F_{xy}^2)/ \sigma^2$$

This plane stress problem is a nice application of optimality criteria methods as can be seen in the following three FORTRAN programs on the CD:

- Prog21.for: This program is a grid generator for a rectangular thin plate. It includes node loads, in this case a unit load at the end of the member as indicated in Figure 4.7.
- Prog20.for: This is a plane stress program that uses the Zienkiewicz constant stress element. It is included here for purposes of completeness.
- Prog22.for: This program is simply FEMPS modified for optimization using the Mises yield condition to generate the appropriate stiffness matrix.

These programs are run in two steps, for example,

Prog21 out

Prog22 out out1

where the output of the first program becomes the input to the second.

In the spirit of the truss program listed in this chapter, Prog22.for is listed here: (1) the finite element program is made to iterate, (2) the member stiffness matrix is modified as described above, and (3) the thickness is modified at each iteration.

```
C     PROG22.FOR. PLANE STRESS FINITE ELEMENT PROGRAM
C     OPTIMIZATION USING ITERATIVE DESIGN
      DIMENSION NP(200,3),B(3),CZ(3),AZ(3),AY(3),S(200)
     1 ,sig(200,3)
      DOUBLE PRECISION R(200),P(200),C(200,200),F1,F2
     1, psave(200)
      E=29.E6
      fy=36000.
      ANU=.3
      E1N=E/(1.-ANU*ANU)
      ON2=.5-ANU/2.
  100 READ(50,150) NB,NN,NS
      NNN=NN-NS
      READ (50,156) (R(2*I-1),R(2*I),Psave(2*I-1),
     1 Psave(2*I),I=1,NN)
      WRITE (60,157)
      WRITE (60,158) (I,R(2*I-1),R(2*I),Psave(2*I-1),
     1 Psave(2*I),I=1,NN)
      N=2*NN
C
C     SET UP SYSTEM MATRIX
      nit=40
      do 9999 iter=1,nit
      volume=0.
      DO 30 I=1,N
      p(i)=psave(i)
      DO 30 J=1,N
```

```
   30 C(I,J)=0.
      N=2*NNN
      WRITE (60,159)
      L11=0
      L=0
  201 L=L+1
      if(iter.eq.1)
     1 READ(50,151) (NP(L,I),I=1,3),S(L)
      WRITE(60,160) L,(NP(L,I),I=1,3),S(L)
      call bc(np,r,b,cz,area,l)
  104 DO 13 L1=1,3
      IR=2*NP(L,L1)
      IF(IR.GT.N) GO TO 13
      DO 14 L2=1,3
      IS=2*NP(L,L2)
      IF(IS.GT.N) GO TO 14
      if(iter.ne.1) go to 5555
      Z11=E1N*S(L)*.25/AREA
      C(IR-1,IS-1)=C(IR-1,IS-1)+Z11*B(L1)*B(L2)
     1 +Z11*ON2*CZ(L1)*CZ(L2)
      C(IR-1,IS  )=C(IR-1,IS  )+Z11*(ANU*B(L1)*CZ(L2)
     1 +ON2*B(L2)*CZ(L1))
      C(IR,IS-1)=C(IR,IS-1)+Z11*ON2*B(L1)*CZ(L2)
     1 +Z11*ANU*B(L2)*CZ(L1)
      C(IR,IS)=C(IR,IS)+Z11*(CZ(L1)*CZ(L2)+ON2*B(L1)*B(L2))
      go to 14
 5555 Z11=2.*fy*fy*S(L)*.25/AREA
      C(IR-1,IS-1)=C(IR-1,IS-1)+Z11*B(L1)*B(L2)*.6667
     1 +Z11*CZ(L1)*CZ(L2)/6.
      C(IR-1,IS  )=C(IR-1,IS  )+Z11*(B(L1)*CZ(L2)/3.
     1 +B(L2)*CZ(L1)/6.)
      C(IR,IS-1)=C(IR,IS-1)+Z11*B(L1)*CZ(L2)/6.
     1 +Z11*B(L2)*CZ(L1)/3.
      C(IR,IS)=C(IR,IS)+Z11*(CZ(L1)*CZ(L2)*.6667+B(L1)*B(L2)/6.)
   14 CONTINUE
   13 CONTINUE
   12 IF(L-NB) 201,202,202
C
C     SOLVE THE SYSTEM
  202 M=N-1
      DO 17 I=1,M
      L=I+1
      DO 17 J=L,N
      IF (C(J,I))19,17,19
   19  DO 18 K=L,N
   18 C(J,K)=C(J,K)-C(I,K)*C(J,I)/C(I,I)
      P(J)=P(J)-P(I)           *C(J,I)/C(I,I)
   17 CONTINUE
      P(N)=P(N)/C(N,N)
      DO 20 I=1,M
      K=N-I
      L=K+1
      DO 21 J=L,N
   21 P(K)=P(K)-P(J)*C(K,J)
      P(K)=P(K)/C(K,K)
   20 CONTINUE
C
C     OUTPUT DISPLACEMENTS
      WRITE (60,161)
```

```
      WRITE (60,152)(I,P(2*I-1),P(2*I),I=1,NNN)
      WRITE (60,162)
      L11=1
      L=0
  203 L=L+1
      call bc(np,r,b,cz,area,l)
  102 DO 204 K=1,3
  204 AZ(K)=0.
      DO 34 L1=1,3
      IR=2*NP(L,L1)
      IF(IR.GT.N) GO TO 34
      AZ(1)=AZ(1)+B(L1)*P(IR-1)*.5/AREA
      AZ(2)=AZ(2)+CZ(L1)*P(IR)*.5/AREA
      AZ(3)=AZ(3)+(CZ(L1)*P(IR-1)+B(L1)*P(IR))*.5/AREA
   34 CONTINUE
      if(iter.ne.1)e1n=2.*fy*fy
      AY(1)=E1N*(AZ(1)*.6667+.3333*AZ(2))
      AY(2)=E1N*(.3333*AZ(1)+.6667*AZ(2))
      AY(3)=E1N*AZ(3)/6.
      WRITE(60,952) L,(AZ(I),I=1,3),(AY(I),I=1,3)
      do 567 k=1,3
  567 sig(1,k)=ay(k)
      call mises(sig,s,fy,1)
      volume=volume+area*s(1)
   22 IF(L-NB) 203,888,888
  888 continue
      write(60,*) 'iteration',iter,'volume',volume
 9999 continue
      stop
  150 FORMAT(3(I4,3X))
  151 FORMAT(3I5,8X,E20.8)
  156 FORMAT(8X,4F11.6)
  152 FORMAT(I10,2D20.8)
  952 FORMAT(I10,6D20.8)
  157 FORMAT(1H1,18X,11HCOORDINATES,32X,5HLOADS/
     114X,1HX,19X,1HY,18X,2HPX,18X,2HPY//)
  158 FORMAT(I4,4D20.8)
  159 FORMAT(1H1,2X,7HELEMENT,13X,13HELEMENT NODES,15X,9HTHICKNESS//)
  160 FORMAT(4I10,E20.8)
  161 FORMAT(1H1,13HDISPLACEMENTS/20X,1HX,19X,1HY//)
  162 FORMAT(1H1,35X,7HSTRAINS,52X,8HSTRESSES//3X,7HELEMENT,
     1 9X,2HEX,18X,2HEY,18X,2HGA,18X,2HSX,18X,2HSY,18X,3HTAU//)
      END
c
      subroutine bc(np,r,b,cz,area,l)
      dimension np(200,3),b(1),cz(1)
      double precision r(1)
  101 I=NP(L,1)
      J=NP(L,2)
      M=NP(L,3)
      B(1)=-R(2*M)+R(2*J)
      B(2)=R(2*M)-R(2*I)
      B(3)=-R(2*J)+R(2*I)
      CZ(1)=R(2*M-1)-R(2*J-1)
      CZ(2)=-R(2*M-1)+R(2*I-1)
      CZ(3)=R(2*J-1)-R(2*I-1)
      AREA=.5*(R(2*I)*CZ(1)+R(2*J)*CZ(2)+R(2*M)*CZ(3))
      return
      end
```

```
c
      subroutine mises(sig,s,fy,l)
      dimension s(1), sig(200,3)
      phi=sqrt(sig(l,1)**2-sig(l,1)*sig(l,2)+sig(l,2)**2
     1 +3.*sig(l,3)**2)
      s(l)=s(l)*phi/fy
      return
      end
```

4.11 Prager and His Co-Workers

Some of the very early work in structural optimization was carried out by the late William Prager and his co-workers at Brown University. That work is described in some detail in Save and Prager (1985). It reflects the very strong mechanics program at Brown University at the time the work was done. The work discussed below is plastic design and it is essentially a generalization of fully stressed design for trusses.

4.11.1 The Axisymmetric Circular Plate

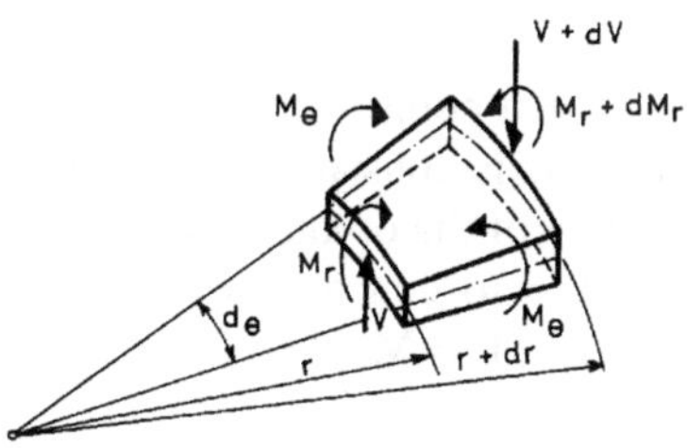

Fig. 4.8. A Plate Element

Probably the simplest problem (Onat et al., 1957 and Shield, 1959) of the mechanics type (beyond the beam) is the axisymmetric circular, sandwich plate (Fig. 4.8). The problem is, given some normal load w, find the thickness t(r) that will satisfy some failure criterion and minimize the total weight of the plate. We will look at the case here of a sandwich plate under a uniformly distributed load that satisfies the Tresca yield condition (Fig. 4.9).

For the axisymmetric plate the equilibrium equations are

$w\,r + V + r\,dV/dr = 0$ (vertical equilibrium)

$V = (M_R - M_\Theta)/r + dM_R/dr$ (moment equilibrium)

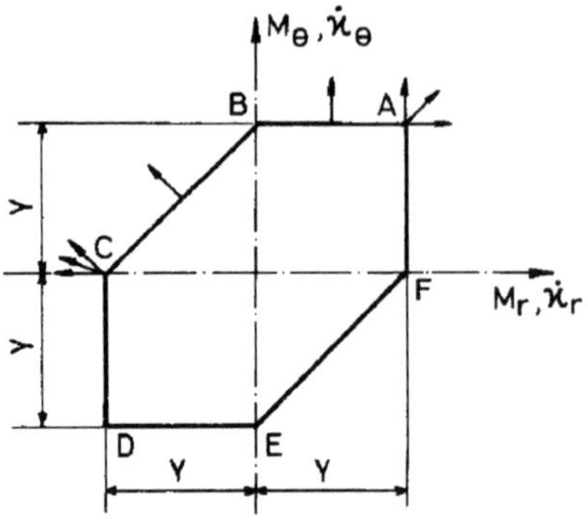

Fig. 4.9. Tresca Yield Condition (in terms of moments)

Keeping with the notation of the node method, the member forces and displacements are

$$F = \begin{bmatrix} M_R \\ M_\Theta \end{bmatrix} \qquad \textit{and} \qquad \Delta = \begin{bmatrix} k_R \\ k_\Theta \end{bmatrix} = \begin{bmatrix} -d^2u_R/dr^2 \\ (-du_R/dr)/r \end{bmatrix}$$

with u_R the plate displacement. Prager now argues that point A on the Tresca yield surface must correspond to the optimal design. This is consistent with the above discussion of the dual problem in which the product of the optimal member forces and displacements must be maximal. (Prager likes to couch this discussion in terms of the *rate of energy dissipation.*) At point A, $M_R = M_\Theta$ and the equilibrium equations give

$$rd^2M_R/dr + dM_R/dr = -rw$$

or

$$M_R = -r^2w/4 - c_1/r^2 + c_2$$

For the solid plate (a plate with no center opening), c_1 must be zero. At the support $r = R$ the moment must be zero. It follows that

$$M_R = w(R^2 - r^2)/4$$

Finally, given the load w, and a sandwich plate where $M_R = t\,\sigma_{yield}\,h$ with

t – flange plate thickness
σ_{yield} – material yield stress
h – core thickness of the sandwich plate

the optimal thickness has been determined. The reader is encouraged to look at Prager's discussion of more complex cases.

4.12 A Plate Problem

As a final problem for this chapter we return to Section 4.6 and consider the design of a square (this time), fixed sandwich plate under a uniform load. This solution again uses the method of finite differences and conditions of symmetry. Symmetry allows a ¼ model to be used. (Actually, a 1/8 model could have been used but the 1/8 model seemed more difficult to the authors.) We use the common theory of bending of plates and use the Mises failure theory to calculate the plate flange thickness given the bending moments. In passing, we looked at the failure model used by Rozvany and discussed in a later chapter. Rozvany's model is built around the principal moments and we found it to be difficult to use because of the inherent discontinuities it possesses in its derivatives. The following figures (Figs. 4.10 and 4.11) show both the optimal flange thickness and the optimal flange plate thickness.

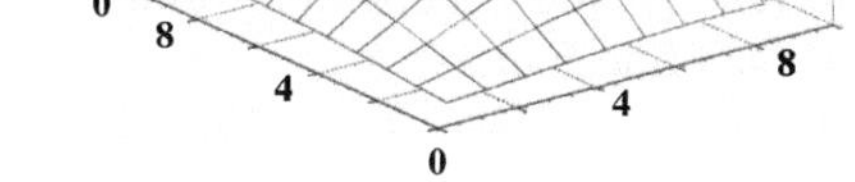

Fig. 4.10. Optimal Displacement Field

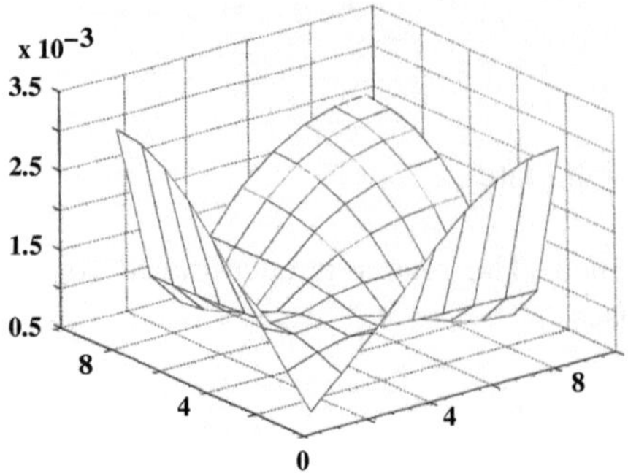

Fig. 4.11. Optimal Flange Plate Thickness

The formulation begins with the plate element (Fig. 4.12) showing the bending moments, the displacement $y(x,y)$, the generalized strains Δ, and the stiffness matrix K. These are written in the notation of the node method that was introduced earlier

$$F = \begin{bmatrix} M_x \\ M_y \\ M_{xy} \end{bmatrix}, \quad \Delta = \begin{bmatrix} -y_{,xx} \\ -y_{,yy} \\ 2y_{,xy} \end{bmatrix}, \quad and \quad K = \frac{Et^3}{12(1-\nu^2)} \begin{bmatrix} 1 & \nu & \\ \nu & 1 & \\ & & \frac{1-\nu}{2} \end{bmatrix}$$

K in this case is the elastic stiffness matrix. The moment equations of equilibrium are in this case

$$M_{x,xx} + M_{y,yy} - 2M_{xy,xy} = -q \qquad or\ simply \qquad LF = -q$$

with q the distributed load and L the equilibrium equation in operator form.

The *optimization* problem to be solved is then find F to

$$minimize \int_S T(M)dS \qquad subject\ to \qquad LF = -q \quad \text{(Equilibrium)}$$

with

$$T(M) = \sqrt{M_x^2 - M_x M_y + M_y^2 + 3M_{xy}^2}$$

using the Mises yield condition. This is, of course, again plastic design.

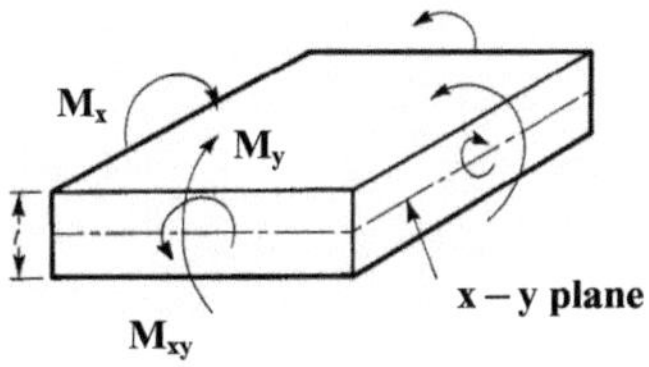

Fig. 4.12. A Plate Element

Following the method described earlier in this chapter, an iterative solution is used in which the stiffness matrix comes from the objective function, in this case, T. Here T is again homogeneous of degree 1. Let $\Phi = T^2$ and H be the Hessian matrix of Φ. The iterative process then uses a new stiffness matrix

$$K^{-1} = H / \sqrt{4\Phi}$$

Here

$$\Phi = T^2, \quad \nabla\Phi = \begin{bmatrix} 2M_x - M_y \\ -M_x + 2M_y \\ 6M_{xy} \end{bmatrix}, \quad and \quad H = \begin{bmatrix} 2 & -1 & 0 \\ -1 & 2 & 0 \\ 0 & 0 & 6 \end{bmatrix}$$

K follows directly.

The finite difference solution again divides the plate into a two-dimensional grid. For the iterative solution, the equation to be solved is

$$K_{11}\, y_{,xxxx} + K_{22}\, y_{,yyyy} + (2\, K_{12} + 4\, K_{33})\, y_{,xxyy} + q = 0$$

where the K's are elements of the above stiffness matrix. This solution requires two finite difference operators, the fourth derivative $y_{,xxxx} = 1/h^4$ x

(1)—(-4)—(6)—(-4)—(1)

and the mixed order fourth derivative $y_{,xxyy} = 1/h^4$ x

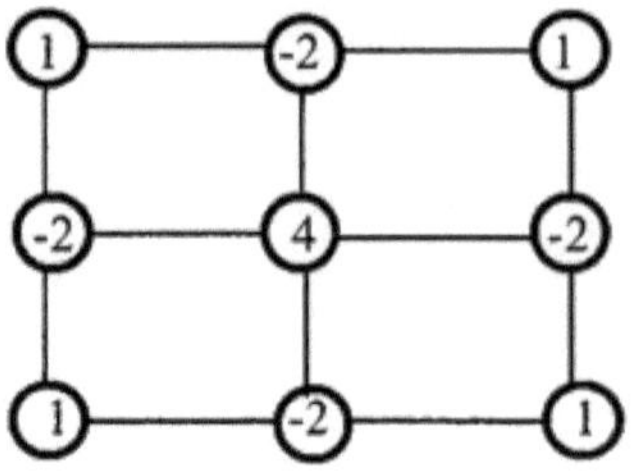

The solution to this problem is included on the CD as the program PProg23.for. This program is typical of the iterative design process in which an *elastic-type* analysis is performed, the stiffness is then modified using the analysis just performed, and another analysis is performed, etc.

4.13 Problems

1. Linear programming and the absolute value function. Derive the dual truss design problem using the traditional linear programming dual.

   Solution. The truss problem of Section 4.1 is written in terms of the absolute value of the member forces, F_i. It is common in the linear programming literature to treat the absolute value using the following transformation:

$$F_i = F_i^+ - F_i^- \quad with \quad F_i^+, F_i^- \geq 0 \quad and \quad |F_i| = F_i^+ + F_i^+$$

   The truss problem

$$\text{minimize } |F| \cdot \Delta_a \qquad \text{subject to} \qquad N^T F = P$$

   then becomes

$$\text{minimize } |F^+ + F^-| \cdot \Delta_a \qquad \text{subject to} \qquad N^T(F^+ - F^-) = P$$

Using the standard form of linear programming problem
Primal problem:

$$minimize\ c^T x \qquad subject\ \ to \qquad Ax = b, \quad x \geq 0$$

Dual Problem:

$$maximize\ b^T y \qquad subject\ \ to \qquad A^T y \leq c$$

to show that the dual system for the truss problem is
Primal problem:

$$\textit{minimize } |F| \cdot \Delta_a \qquad \textit{subject to} \qquad N^T F = P$$

Dual problem:

$$\textit{maximize } P^T \delta \qquad \textit{subject to} \qquad |\Delta| = |N\delta| \leq \Delta_a$$

Solution. Let

$$F^* = \begin{bmatrix} F^+ \\ F^- \end{bmatrix} \qquad \textit{then} \qquad N^T F = P \;\Rightarrow\; (N^*)^T F^* = P$$

$$\textit{with} \qquad N^* = \begin{bmatrix} N \\ -N \end{bmatrix}$$

The structures problem then translates into find F^* to

$$\textit{minimize } F^* \cdot \Delta_a^* \quad \textit{subject to} \quad (N^*)^T F^* = P, \; F^* \geq 0$$

It only remains to do the transformation of the dual problem.

2. Redo the truss problem of this chapter so that it reflects the allowable stress in compression used by the AISC or some European code.

3. Extend the formulation of the truss problem to reflect shape change.
4. Modify the plane stress program FEMOPT and solve for the effect of a hole or several holes.

5 Some Basic Optimization Problems

This chapter discusses some basic optimization results from the areas of multiple loading conditions, deflection constraints, optimal shape, and design automation.

5.1 Multiple Loading Conditions

There is an interesting result for the truss problem that states that design for two loading conditions can be treated as the sum of two single loading condition designs. It seems to be due to Hemp (1968) and rediscovered by Lev and Spillers (1971). It again treats the plastic design of trusses or an embedded elastic design with constant allowable stresses. The embedded design raises the issue of realizability and an example will be discussed that is not realizable.

The plastic design problem for two loading conditions can be written as

Primal Problem

$$\textit{minimize } \phi = \frac{\sigma}{E}\sum_i L_i \max\{|F_i^1|,|F_i^2|\}$$

$$\textit{subject to } N^T F^1 = P^1 \textit{ and } N^T F^2 = P^2$$

Dual Problem

$$\textit{maximize } \psi = (P^1)^T \delta^1 + (P^2)^T \delta^2$$

$$\textit{subject to } |N\delta^1| + |N\delta^2| \leq \frac{\sigma L}{E}$$

where the superscripts refer to the loading condition. The primal problem here is obvious to the extent that an attempt is made to treat the worst-case bar forces. If the dual is not obvious, it can be obtained as a standard linear programming dual. It is included here more for reasons of completeness.

In order to decompose this problem, two identities are required:

$$\max\{|x|,|y|\} = \frac{1}{2}|x+y| + \frac{1}{2}|x-y|$$

and

$$|x| + |y| \leq 1 \Leftrightarrow |x+y| \leq 1 \quad \textit{and} \quad |x-y| \leq 1$$

W.R. Spillers, K.M. MacBain, *Structural Optimization*,
DOI 10.1007/978-0-387-95865-1_5, © Springer Science+Business Media, LLC 2009

These identities can be demonstrated by exhaustive trial and error. When they are inserted into the dual programming problem, it follows that

Primal Problem

$$\textit{minimize}\ \phi = \frac{\sigma}{E}\sum_i L_i \ \{|\ F_i^1 + F_i^2\ |\ /\ 2 + |\ F_i^1 - F_i^2\ |\ /\ 2\}$$

$$\textit{subject to}\ \ N^T F^1 = P^1 \ \textit{and}\ N^T F^2 = P^2$$

Dual Problem

$$\textit{maximize}\ \psi = (P^1)^T \delta^1 + (P^2)^T \delta^2$$

$$\textit{subject to}\ \ |\ N\delta^1 + N\delta^2\ | \le \sigma L / E \quad \textit{and} \quad |\ N\delta^1 - N\delta^2\ | \le \sigma L / E$$

which can be rewritten as

Primal Problem

$$\textit{minimize}\ \phi = \frac{\sigma}{E}\sum_i L_i \ \{|\ F_i^1 + F_i^2\ |\ /\ 2 + |\ F_i^1 - F_i^2\ |\ /\ 2\}$$

$$\textit{subject to}\ \ N^T\left(\frac{F^1 + F^2}{2}\right) = \frac{P^1 + P^2}{2} \quad \textit{and}$$

$$N^T\left(\frac{F^1 - F^2}{2}\right) = \frac{P^1 - P^2}{2}$$

Dual Problem

$$\textit{maximize}\ \psi = \left(\frac{P^1 + P^2}{2}\right)^T (\delta^1 + \delta^2) + \left(\frac{P^1 - P^2}{2}\right)^T (\delta^1 - \delta^2)$$

$$\textit{subject to}\ \ |\ N(\delta^1 + \delta^2)\ | \le \sigma L / E \quad \textit{and} \quad |\ N(\delta^1 - \delta^2)\ | \le \sigma L / E$$

It is now convenient to introduce new *sum* and *difference* variables

$F^S = ½(F^1 + F^2)$ $\qquad$ $F^D = ½(F^1 - F^2)$

$P^S = ½(P^1 + P^2)$ $\qquad$ $P^D = ½(P^1 - P^2)$

$\delta^S = \delta^1 + \delta^2$ $\qquad$ $\delta^D = \delta^1 - \delta^2$

Since the sum and difference variables are independent, the coupled dual problem decomposes into two uncoupled problems in terms of the new variables:

Sum Problem

$$\text{Minimize} \quad \phi^S = \frac{\sigma}{E}\sum_i L_i \mid F_i^S \mid \quad \text{subject to} \quad N^T F^S = P^S$$

$$\text{Maximize} \quad \psi^S = (P^S)^T \delta^S \quad \text{subject to} \quad \mid N\delta^S \mid \leq \sigma L / E$$

Difference Problem

$$\text{Minimize} \quad \phi^D = \frac{\sigma}{E}\sum_i L_i \mid F_i^D \mid \quad \text{subject to} \quad N^T F^D = P^D$$

$$\text{Maximize} \quad \psi^D = (P^D)^T \delta^D \quad \text{subject to} \quad \mid N\delta^D \mid \leq \sigma L / E$$

Since both the sum and difference problems have the form of single loading condition problems, the methods developed earlier apply directly to the case of two loading conditions. Note that the 25-bar truss problem of Chapter 3 begins with an application of this decomposition.

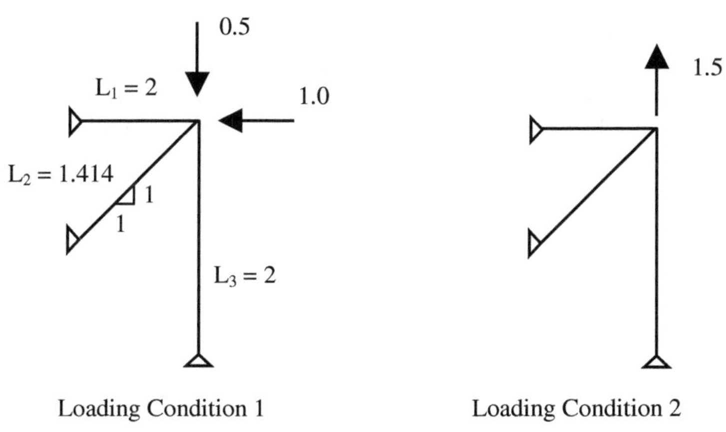

Fig. 5.1. A Problem with Two Loading Conditions

5.1.1 An Example

Figure 5.1 shows a simple three-bar truss. The optimal forces in this case are

$$F^S = \begin{bmatrix} -1/2 \\ 0 \\ 1/2 \end{bmatrix} \qquad F^D = \begin{bmatrix} 0 \\ -1/\sqrt{2} \\ -1/2 \end{bmatrix}$$

When E = σ = 1, the optimal areas are

$$A_1 = 1/2 \qquad A_2 = 1/\sqrt{2} \qquad A_3 = 1$$

for an optimal volume of 3 1/2.

5.1.2 Realizability

Plastic design or embedding raises the issue of realizability. That is, once a force system (which implies member areas) has been determined, is it possible to construct an elastic solution that will return this system of forces? In this case the answer is no. But it may be noted that when an optimal design is statically determinate it is generally possible to realize the design. In the example shown above, while the sum and difference solutions are statically determinate, when they are combined or summed, the design is no longer statically determinate.

Rather than simply attempting elastic realizability, realizability using prestress will be considered. In the case of prestress, the equations of structures can be written as follows:

$$\begin{aligned} N^T F &= P && (\textit{equilibrium}) \\ F &= K(\Delta - D) && (\textit{constitutive equations}) \\ \Delta &= N\delta && (\textit{member / joint displacement equations}) \end{aligned}$$

Here the prestress forces are written as KD with D_i the member length change associated with the prestress of bar i.

In discussing realizability it is convenient to work with the load *sequence*

Load P^1 → Load $(P^2 - P^1)$ → Load P^2

The point is that the prestress does not effect the response to the Load $(P^2 - P^1)$. Some analysis for this problem shows that

$$N = \begin{bmatrix} 1 & 0 \\ 1/\sqrt{2} & 1/\sqrt{2} \\ 0 & 1 \end{bmatrix} \qquad K = \frac{1}{2}\begin{bmatrix} 1 & 0 & 0 \\ 0 & 1 & 0 \\ 0 & 0 & 1 \end{bmatrix}$$

$$N^T KN = \frac{1}{4}\begin{bmatrix} 3 & 1 \\ 1 & 3 \end{bmatrix} \qquad (N^T KN)^{-1} = \frac{1}{2}\begin{bmatrix} 3 & -1 \\ -1 & 3 \end{bmatrix}$$

Applying the loads $P^2 - P^1$ produces the following changes in member forces:

$$F = KN(N^T KN)^{-1}\begin{bmatrix} 1 \\ 2 \end{bmatrix} = \begin{bmatrix} 1/4 \\ \sqrt{2}\,3/4 \\ 5/4 \end{bmatrix}$$

If the structure is to be safe under P^2, the load in member 3 must be less than or equal to 1 since member 3 has unit area. Setting this force equal to 1 gives the bar forces as

$$F = \begin{bmatrix} -1/2 \\ 1/\sqrt{2} \\ 1 \end{bmatrix}$$

This implies that the bar forces under P^1 are

$$F = \begin{bmatrix} -3/4 \\ -1/4\sqrt{2} \\ -1/4 \end{bmatrix}$$

The dilemma is as follows: Bar 1 is overstressed under P^1 since it has an area of ½ but any attempt to relieve its load increases the load in bar 3 under P^2. The conclusion is that the optimal force system is not realizable.

It is of interest to note that the problem of elastic realizability is identical with the *shakedown* question of plastic analysis discussed by Symonds and Prager (1950). That is, a set of loads which shakedown are by definition elastically realizable.

5.1.3 Three Loading Conditions

It will be seen that problems of three (or more) loading conditions do not decompose as the two loading condition problem does. For the case of three loading conditions, the dual problem is

Primal Problem

$$\text{minimize} \quad \varphi = \frac{\sigma}{E}\sum_i L_i \max\left[|F_i^1|, |F_i^2|, |F_i^3|\right]$$

$$\text{subject to} \quad N^T F^1 = P^1, \; N^T F^2 = P^2, \; N^T F^3 = P^3$$

Dual Problem

$$\text{maximize} \quad \psi = (P^1)^T \delta^1 + (P^2)^T \delta^2 + (P^3)^T \delta^3$$

$$\text{subject to} \quad |N\delta^1| + |N\delta^2| + |N\delta^3| \leq \sigma L / E$$

Probably the most direct way to approach the decomposition is through the constraints of the dual. For the case of two loading conditions it was noted that

$$|x| + |y| \leq 1 \quad \Leftrightarrow \quad \begin{array}{c} |x+y| \leq 1 \\ \\ |x-y| \leq 1 \end{array}$$

from which the decomposition follows directly. Figure 5.2 shows this inequality graphically and indicates that the shaded region can be described in terms of four lines. For the case of three loading conditions this planar region generalizes to an octagon as indicated in this figure. This region can be described in terms of eight planes and implies the following identity:

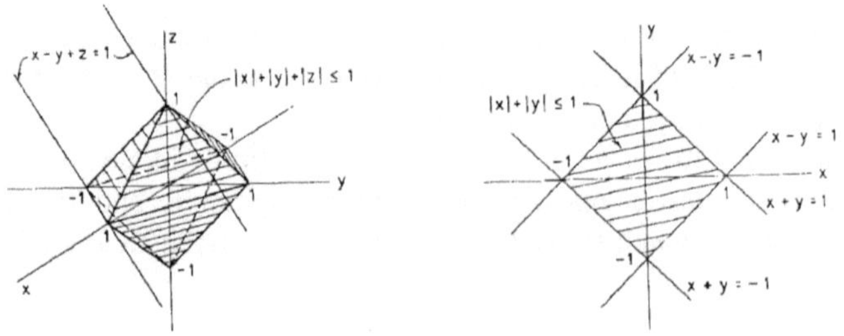

Fig. 5.2. Constraints of the Dual

$$
|x|+|y|+|z|\le 1 \quad \Leftrightarrow \quad
\begin{array}{l}
|x+y+z|\le 1 \\
|-x+y+z|\le 1 \\
\\
|-x-y+z|\le 1 \\
|x-y+z|\le 1
\end{array}
$$

This motivates writing the dual problem for three loading conditions as

Dual Problem

$$
\begin{aligned}
&\textit{maximize} \quad \psi=(p^1)^T d^1+(p^2)^T d^2+(p^3)^T d^3+(p^4)^T d^4 \\
&\textit{subject to} \quad |Nd^1|\le \sigma L/E, \quad |Nd^2|\le \sigma L/E, \\
&\qquad\qquad\quad |Nd^3|\le \sigma L/E, \quad |Nd^4|\le \sigma L/E
\end{aligned}
$$

in which

$$
\begin{array}{ll}
p^1=(P^1+P^2+P^3)/4 & d^1=\delta^1+\delta^2+\delta^3 \\
p^2=(-P^1+P^2+P^3)/4 & d^2=-\delta^1+\delta^2+\delta^3 \\
p^3=(-P^1+-P^2+P^3)/4 & d^3=-\delta^1-\delta^2+\delta^3 \\
p^4=(P^1-P^2+P^3)/4 & d^4=\delta^1-\delta^2+\delta^3
\end{array}
$$

Since neither the p's nor the d's are independent, the three loading condition problem does not decompose like the two loading condition problem does.

The case of n loading conditions is now clear. The feasible region of Fig. 5.2 requires 2^n planes for its definition so that the right-hand side of the basic identity in general contains ½ (2^n) terms. It is therefore fairly obvious that the decomposition does not hold for n > 2 and that algebraic difficulties compound themselves quickly with increasing n.

5.2 Deflection Constraints

It is common to be concerned with the deflection of structures. These concerns could have to do with the functioning of equipment, comfort of the occupants of a building under wind load, the motion due to pedestrian traffic in a building or on a bridge, etc. Or you may simply want to stiffen a structure and do so in an *optimal* manner. This section considers some of the more elementary issues regarding deflection constraints. We will return to some more complicated issues in a later chapter.

Figure 5.3 illustrates a rather surprising feature of deflection constraints.

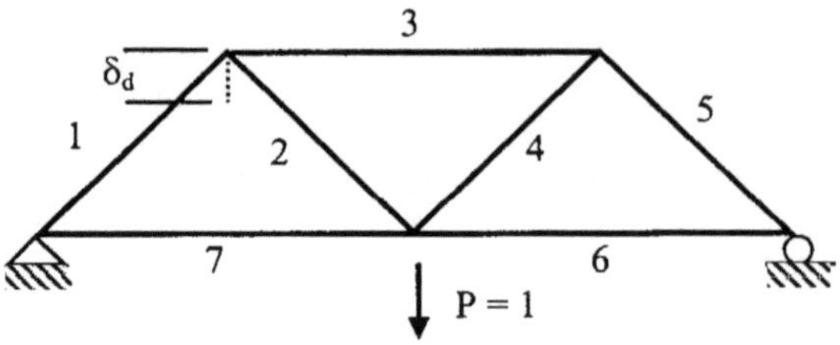

Fig. 5.3. A Simple Truss Problem

In this case a statically determinate truss is given a single displacement constraint δ_d. Using the method of virtual work this constraint can be written as

$$\sum \frac{F_i f_i L_i}{A_i E} = \delta_d$$

where

δ_d – displacement constraint
F_i – (real) force in bar i
f_i – (virtual) force in bar i
L_i – length of bar i
E – Young's modulus

When this problem is returned to it will be seen that by making the area of bar 2 small, both the displacement δ_d and the volume can be made as small as you like. (It will be seen that this bar area must be controlled by a stress constraint.)

Figure 5.4 describes a truss problem in which it is desired to minimize the volume for a given displacement at the top. (Hence the loading of the virtual structure.) The problem statement is then find the bar areas A_i to

$$\textit{minimize} \quad \sum A_i L_i \qquad \textit{subject to} \qquad \sum \frac{F_i f_i L_i}{A_i E} = d$$

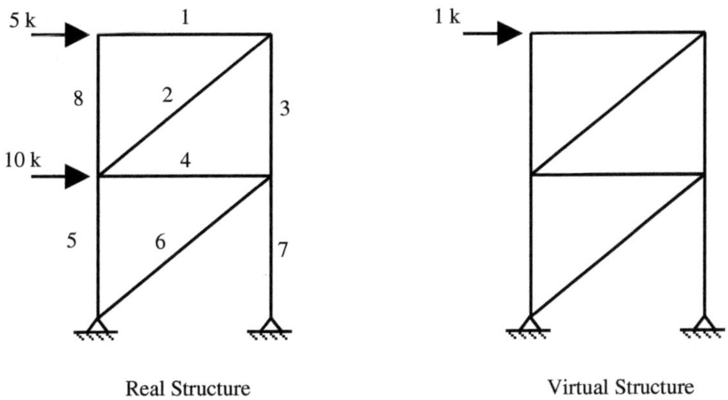

Fig. 5.4. A Truss with a Displacement Constraint

Using the Lagrange multiplier method, the Lagrangian is formed and differentiated with regard to the unknown areas,

$$L=\sum A_i L_i+\lambda\left(-d+\sum\frac{F_i f_i L_i}{A_i E}\right),\quad \partial L/\partial A_i=0\ \Rightarrow\quad L_i+\lambda\frac{F_i f_i L_i}{A_i^2 E}=0$$

Clearly this equation specifies the ratio of the bar areas

$$A_i=\sqrt{\lambda\frac{F_i f_i}{E}}$$

after which the deflection constraint can be used to compute λ. This is illustrated in Table 5.1, where bar 8 has been excluded because it does not contribute to the computations.

TABLE 5.1 Virtual Work Computations

Bar	F_i	f_i	L_i	$\sqrt{(F_i f_i)}$	$L_i\sqrt{(F_i f_i)}$	A_i
1	−5	−1	10	√5	10 √5	0.086
2	5√2	√2	10√2	√10	10 √20	0.1228
3	−5	−1	10	√5	10 √5	0.086
4	−15	−1	10	√15	10 √5	0.1504
5	5	1	10	√5	10 √5	0.086
6	15√2	√2	10√2	√30	10 √60	0.2127
7	−20	−2	10	√40	10 √40	0.2456

If $d = 0.25''$ and $E = 30{\times}10^3$ ksi, then $\lambda = 6.726^2$ and the areas follow directly.

This is the most simple type of problem with a deflection constraint: The Lagrange multiplier implies the ratio of the bar areas and the deflection constraint determines the values of the areas.

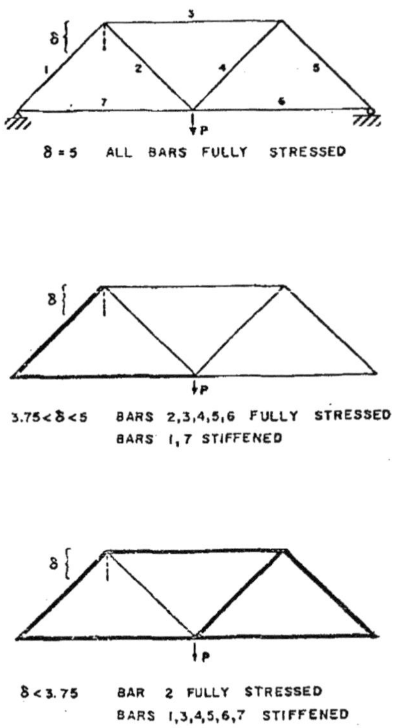

Fig. 5.5. Areas Controlled by the Value of the Displacement Constraint

5.2.1 Funaro's Example

Figure 5.3 describes a more complex example. This example is due to Funaro (1974). It is a problem in which there are both deflection and stress constraints. As indicated in Fig. 5.5 the optimal design depends on the value of the displacement constraint. That is, as you make the displacement constraint tighter and tighter, you go from a design controlled by stress constraints to one in which different sets of bars are *stiffened* or controlled by the value of the displacement constraint.

This example includes two new effects. One is that the allowable stress constraints are added and the other is the fact that the virtual force and the real force in bar 2 have different signs (see Table 5.2). Dealing with the latter first, in view of the form of the virtual work equation, the area of bar 2 should be made as small as possible so that this bar is always stressed to the allowable no matter what the displacement constraint.

TABLE 5.2 Virtual Work Computations

Bar	L_i	F_i	f_i	f/F_i	$\lvert f_i\rvert L_i$	$L_i\sqrt{(F_i f_i)}$
1	1.414	–0.707	–1.06	1.5	1.5	1.224
2	1.414	0.707	–0.354	–0.5	0.5	–
3	2	–1	–0.5	0.5	1.0	1.414
4	1.414	0.707	0.354	0.5	0.5	0.707
5	1.414	–0.707	–0.354	0.5	0.5	0.707
6	2	0.5	0.25	0.5	0.5	0.707
7	2	0.5	0.75	1.5	1.5	1.224

The formulation of this problem, with the allowable stress constraints, is find A_i to

$$\textit{minimize}\ \sum A_i L_i \qquad \textit{subject to} \qquad \sum \frac{F_i f_i L_i}{A_i E} \le d \ \textit{ and } \ \frac{|F_i|}{A_i} \le \sigma_{allow}$$

Funaro's procedure goes as follows. He first forms the Lagrangian

$$L = -\sum A_i L_i + \lambda\left(d - \sum \frac{F_i f_i L_i}{A_i E}\right) + \sum \mu_i \left(\sigma_{allow} - |F_i| / A_i\right)$$

Here there are the additional Lagrange multipliers μ_i. The implied *Kuhn–Tucker conditions* are then

$$\partial L / \partial A_i = 0 \qquad \Rightarrow \qquad -L_i + \lambda \frac{F_i f_i L_i}{A_i^2 E} + \mu_i |F_i| A_i^{-2} = 0$$

$$\partial L / \partial \lambda = 0 \qquad \Rightarrow \qquad d = \sum \frac{F_i f_i L_i}{AE}$$

$$\mu_i \left(\sigma_{allow} - |F_i| / A_i\right) = 0$$

Note that the following is the first use in this text of the Kuhn–Tucker conditions. Funaro now defines two sets

B_S – set of all bars for which $|F_i| = A_i\, \sigma_{\text{allow}}$ in the optimal solution
B_D – set of all bars for which $|F_i| < A_i\, \sigma_{\text{allow}}$ in the optimal solution

and shows how they can be determined from the Lagrange multiplier λ. From the Kuhn–Tucker conditions it follows that

$\mu_i = 0$ when $i \in B_D$

and that

$L_i = \lambda\,(F_i f_i L_i / A_i^2 E)$

or that

$$A_i = \frac{\sqrt{\lambda}}{\sqrt{E}}\sqrt{(F_i f_i)} \qquad \textit{for} \qquad i \in B_D$$

Since $A_i \geq |F_i| / \sigma_{allow}$ for bars in this set, it follows that

$$1/\sqrt{\lambda} \leq \frac{\sigma_{allow}}{\sqrt{E}}\sqrt{\frac{f_i}{F_i}}$$

Given λ, this last equation can be used as a test for membership in the set B_D. Since

$$A_i = \sqrt{(\lambda / E)}\sqrt{(F_i f_i)} \qquad i \in B_D$$

and

$$A_i = |F_i| / \sigma_{allow} \qquad i \in B_S$$

It follows that

$$d = \sum_{i \in B_S} \frac{F_i f_i L_i}{A_i E} + \sum_{i \in B_D} \frac{F_i f_i L_i}{A_i E}$$

$$= \frac{\sigma_{allow}}{E} \sum_{i \in B_S} |f_i| \operatorname{sgn}(F_i f_i) L_i + \frac{1}{\sqrt{\lambda E}} \sum_{i \in B_D} L_i \sqrt{F_i f_i}$$

It is now clear that λ implies B_S, B_D, A_i, and d. That is, given λ, it is possible to compute these terms and thus the inverse problem has been solved.

In order to solve for λ, Funaro suggests the following procedure:

1. Compute $\sqrt{(f_i / F_i)}$ and order all bars such that
   $\sqrt{(f_1 / F_1)} \le \sqrt{(f_2 / F_2)} \le \sqrt{(f_3 / F_3)} \cdots$
2. Define $1/\sqrt{\lambda_j} = \frac{\sigma_{allow}}{\sqrt{E}} \sqrt{\frac{f_j}{F_j}}$. Note that for each λ_j there exists a d_j computed as described above. It is clear that $d_1 \le d_2 \le d_3 \cdots$
3. Select an i such that $d_{i-1} \le d \le d_i$.
4. The i just selected implies a λ_i which in turn implies B_S and B_D. λ can now be computed as
   $$\sqrt{\lambda} = \frac{1}{\sqrt{E}} \left\{ \sum_{i \in B_D} L_i \sqrt{F_i f_i} \right\} / \left\{ d - \frac{\sigma_{allow}}{E} \sum_{i \in B_S} |f_i| \operatorname{sgn}(F_i f_i) L_i \right\}$$

The solution is now complete. In computer implementations, it is convenient to modify step 1 such that when the sign of *F*/*f* is negative, the value is taken as some small positive number (0.1/*E* is used for the data below). With this modification, details of step 2 are tabulated below, where 1 indicates that an element belongs to B_D.

i	$\sqrt{F_i / f_i}$	λ_t	d	B1	B2	B3	B4	B5	B6	B7
2	(0.1)	100	0.613	1	1	1	1	1	1	1
3	0.707	2	3.732	1	0	1	1	1	1	1
4	0.707	2	3.732	1	0	1	1	1	1	1
5	0.707	2	3.732	1	0	1	1	1	1	1
6	0.707	2	3.732	1	0	1	1	1	1	1
1	1.225	0.667	5	1	0	0	0	0	0	1
7	1.225	0.667	5	1	0	0	0	0	0	1

The table above is instructive in that it shows how the set membership changes with the value of the displacement constraint. For example, at $\lambda = 2$, B_D consists of bars 1, 3–7 whereas at $\lambda = 0.667$, B_D consists of bars 1 and 7. Following step 3, $i = 1$ is selected from the above table and $\lambda = 1.5$ is computed as per step 4. The resulting bar areas are tabulated below:

mem	Bd	A	σ
1	1	1.061	0.667
2	0	0.707	1.0
3	0	1.000	1.0
4	0	0.707	1.0
5	0	0.707	1.0
6	0	0.500	1.0
7	1	0.750	0.667

An alternate solution will now be presented. With this alternative, the solution begins with $A_i = |F_i| / \sigma_{allow}$. This is, of course, the lightest possible design when displacement constraints are neglected. The corresponding value of the displacement constraint d_0 is then computed. If $d_0 \leq d$ the problem is solved. If it is not, the problem is then to find a $dA_i \geq 0$ to

$$\textit{minimize} \quad \sum dA_i L_i \qquad \textit{subject to} \qquad \sum \frac{F_i f_i L_i}{(A_i + dA_i)E} \leq d$$

This formulation is similar to the first problem considered in this chapter. The optimality condition is then

$$A_i + dA_i = \sqrt{\lambda \frac{F_i f_i}{E}}$$

Proceeding with the solution outlined above, $A_i + dA_i$ can now be computed. If all dA_i turn out to be positive the problem is solved. If some dA_i turn out to be negative, for each of these bars the area is fixed so that allowable stress conditions are satisfied. The analysis is then performed again. When the analysis produces bars for which $dA_i \geq 0$ the problem is solved.

The truss of Fig. 5.2 is used to illustrate this procedure. The calculations are indicated in Table 5.3. We consider the case when

TABLE 5.3 Virtual Work Computations

Bar	A_0	$\sqrt{(F_i f_i)}$	L_i	$L_i\sqrt{(F_i f_i)}$	dA	A
1	0.707	0.8656	1.414	1.223	0.351	1.059
2	0.707	–	1.414	–	–	0.707
3	1	0.707	2	1.414	–0.136	1
4	0.707	0.5	1.414	0.707	–0.096	0.707
5	0.707	0.5	1.414	0.707	–0.096	0.707
6	0.5	0.3535	2	0.707	–0.068	0.5
7	0.5	0.612	2	1.224	0.248	0.748

$d = 4$ which is less than d_0 above and $E = 1$. The virtual work equation

$$d = 4 = \sum \frac{F_i f_i L_i}{A_i E} = \frac{F_2 f_2 L_2}{A_2 E} + \sum_{i \neq 2} L_i \sqrt{(F_i f_i)} \sqrt{\frac{E}{\lambda}}$$

can be solved for $\lambda = 1.493$ which gives the values of *dA*. The other values in the table follow directly. Subsequent iterations converge to the same solution presented earlier.

5.2.2 Optimality Criteria Approach

Note first that there is a close relationship between a stress constraint and a displacement constraint since

$$|\sigma| \leq \sigma_{allow} \quad \Rightarrow \quad |\varepsilon| = |\sigma| / E \leq \sigma_{allow} / E$$

This means that fictitious bars can be used to represent displacement constraints (see Fig. 5.6).

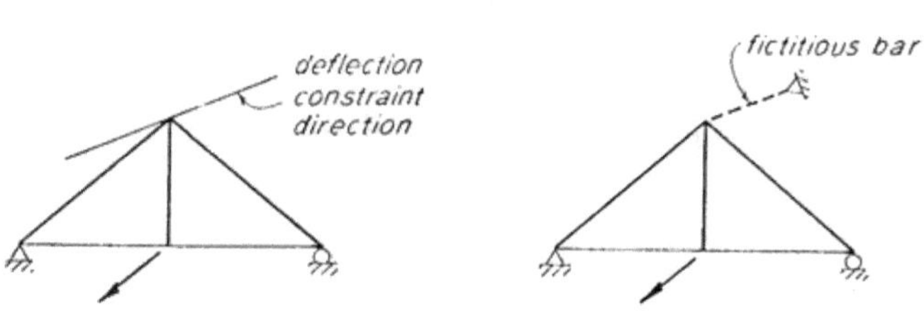

Fig. 5.6. A Displacement Constraint

Using the terminology described earlier in this text, the problem with displacement constraints can now be written as

$$minimize \sum \frac{F_i}{\Delta_i}\left(\Delta_i^{allow}\right)^2$$

$$\begin{aligned} subject\ to\ & N^T F = P \\ & \Delta = N\delta \\ & |\Delta| \le \Delta^{allow} \\ & F_i = 0 \qquad in\ all\ fictitious\ bars \\ & \operatorname{sgn} F_i = \operatorname{sgn} \Delta_i \end{aligned}$$

For the statically determinate case, this problem reduces to

$$minimize \sum \frac{|F_i|}{|\Delta_i|}\left(\Delta_i^{allow}\right)^2$$

$$\begin{aligned} subject\ to\ & N^T F = P \\ & \Delta = N\delta \\ & |\Delta| \le \Delta^{allow} \\ & \operatorname{sgn} F_i = \operatorname{sgn} \Delta_i \end{aligned}$$

In the work that follows, the last constraint will be satisfied by the computational procedure used. The Lagrangian is then

$$L = \sum \frac{|F_i|}{|\Delta_i|}\left(\Delta_i^{allow}\right)^2 + \alpha^T(\Delta - N\delta) + \beta^T\left(\Delta^{allow} - |\Delta|\right)$$

The Kuhn–Tucker conditions for this system are as follows:

$$\begin{aligned} \partial L / \partial \Delta_i = 0 \quad & \Rightarrow \quad -F_i\left(\frac{\Delta_i^{allow}}{\Delta_i}\right)^2 + \alpha_i - \beta_i \operatorname{sgn} \Delta_i = 0 \\ \partial L / \partial \delta_i = 0 \quad & \Rightarrow \quad \left(N^T \alpha\right)_i = 0 \\ \partial L / \partial \alpha_i = 0 \quad & \Rightarrow \quad \Delta_i = (N\delta)_i \\ & \qquad \beta_i\left(\Delta_i^{allow} - |\Delta_i|\right) = 0 \end{aligned}$$

The first two of these conditions can be combined as

$$-N^T\left(F\left(\frac{\Delta^{allow}}{\Delta}\right)^2\right)-N^T(\beta\,\mathrm{sgn}\,\Delta)=0$$

in a somewhat symbolic notation. Let

$$P^* = -N^T F\left(\frac{\Delta^a}{\Delta}\right)^2$$

The Kuhn–Tucker conditions then become

$$\begin{aligned} N^T(\beta\,\mathrm{sgn}\,\Delta) &= P^* \\ \Delta &= N\delta \\ \beta_i(\Delta_i^a - |\Delta_i|) &= 0 \end{aligned}$$

The dual problem has the Lagrangian

$$L = P^T\delta + \gamma^T(\Delta^a - |N\delta|)$$

in which γ is the matrix Lagrange multiplier. The Kuhn–Tucker conditions are

$$\begin{aligned} \partial L/\partial\delta_i = 0 \qquad \Rightarrow \qquad & P_i - (N^T\gamma\,\mathrm{sgn}(N\delta))_i = 0 \\ & \gamma_i(\Delta_i^a - |(N\delta)_i|) = 0 \end{aligned}$$

The similarity of the two sets of Kuhn–Tucker conditions suggests an algorithm. It requires the linearization of the term $(\Delta_i^a/\Delta_i)^2$.

Using a Taylor series approximation,

$$\left(\frac{\Delta_i^{allow}}{\Delta_i}\right)^2 \cong 3\left(\frac{\Delta_i^{allow}}{\Delta_i^{(n)}}\right)^2 - 2\frac{(\Delta_i^{allow})^2}{\left(\Delta_i^{(n)}\right)^3}\Delta_i$$

When multiplied by N^T, the constant term in this equation gives rise to the analog of the joint load term in the node method; the linear term, when written in terms of node displacements, corresponds to a simple modification of the stiffness matrix, something like the geometric term used in nonlinear structural analysis.

Symbolically, this solution procedure involves iterating

$$N^T(K_E^{(n)} + K_G^{(n)})N\delta^{(n+1)} = \overline{P}^{(n)}$$

and

$$(K_E^{(n+1)})_{ii} = (K_E^{(n)})_{ii} \left|\Delta_i^{(n)}\right| / \Delta_i^a$$

$$(K_G^{(n+1)})_{ii} = 2F_i \frac{(\Delta_i^a)^2}{(\Delta_i^{(n)})^3}$$

where

$$\overline{P}^{(n)} = 3N^T(F(\frac{\Delta_i^a}{\Delta_i^{(n)}})^2)$$

For the statically indeterminate case, the Lagrangian becomes

$$L = \sum_i \left|\frac{F_i}{\Delta_i}\right| (\Delta_i^a)^2 + \alpha^T(\Delta - N\delta) + \beta^T(\Delta^a - |\Delta|) + \lambda^T(P - N^T F)$$

This is just the Lagrangian just used with a single additional term. The Kuhn–Tucker conditions then contain two terms

$$(1) \quad \partial L / \partial F_i = 0 \quad \Rightarrow \quad (\operatorname{sgn} F_i)\frac{(\Delta_i^a)^2}{|\Delta_i|} - (N\lambda)_i = 0$$

$$(2) \qquad\qquad N^T F = P$$

in addition to those terms used above.

The general case will be solved iteratively by varying F while holding Δ fixed and then varying Δ while holding F fixed. The case when F is held fixed has been discussed in the preceding section. When Δ is held fixed, the problem to be solved reduces to

$$\textit{minimize} \quad \sum_i \frac{|F_i|}{|\Delta_i|}(\Delta_i^a)^2 \qquad \textit{subject to} \quad N^T F = P$$

This problem is identical in form to the primal problem discussed earlier.

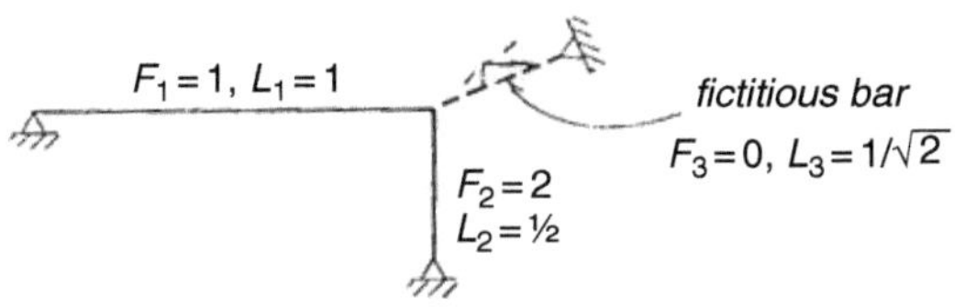

Fig. 5.7. An Example

5.2.3 An Example

Figure 5.7 shows a simple but instructive example. In this case the optimization problem is simply

$$\textit{minimize} \quad \frac{1}{\Delta_1} + \frac{1}{2\Delta_2}$$

subject to $\quad | {}_1| \leq 1$

$\quad | {}_2| \leq ½$

$\quad | {}_1 + {}_2| \leq 1$

$\quad {}_1, {}_2 \geq 0$

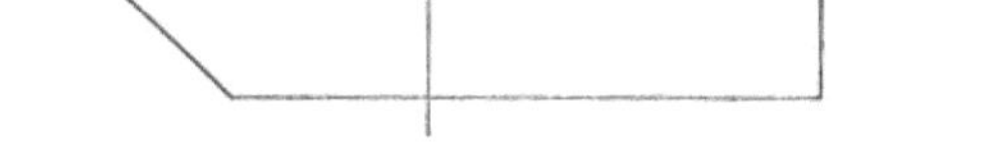

Fig. 5.8. Constraints in Δ Space

Several aspects of this figure are typical of problems with deflection constraints:

1. The solution lies on the boundary of a convex region defined by the linear constraints, but may not be an extreme point. (It is not in this case.)
2. The force vector has been plotted in Fig. 5.8. Since in the absence of deflection constraints the design problem can be written as maximize $P^T\delta = F^T\Delta$,

point A is the optimal solution when the deflection constraint is absent. Point B is easy to get from the linear programming problem, is feasible, and thus a tempting approximate solution to the problem with deflection constraints.

3. The surfaces of constant weight are convex as indicated by Chern and Prager (1971).

5.3 Optimal Shape

This section returns to the work started in Chapter 3 considering the case when structural geometry is allowed to change during the design process. Among the results presented, it will be seen that code design, that includes the effect of member buckling, can produce important changes in the design process.

One of the formulations most used in this text is the problem of finding the member force matrix F to

$$\text{minimize } t(F) \mid \quad N^T F = P$$

with its Lagrangian

$$L = t(F) - \delta^T (P - N^T F)$$

and Kuhn–Tucker conditions

$$\partial L/\partial \delta_i = 0 \qquad \Rightarrow \qquad N^T F = P$$

$$\partial L/\partial Fi = 0 \qquad \Rightarrow \qquad \nabla t = N\delta$$

When the node coordinates are allowed to vary, the Kuhn–Tucker conditions must be augmented to reflect this fact but, of course, all nodes cannot be allowed to move or the structure would end up as a point. Let S be the subset of node coordinates x which are allowed to vary in the optimization procedure. Optimal geometry then requires that for all $x_i \in S$,

$$\partial L/\partial x_i = 0 \quad \Rightarrow \quad \partial t/\partial x_i - \partial\, (\delta^T N^T F)/\partial x_i = 0$$

The terms in this equation will be seen subsequently to depend upon the type of structure under discussion, but the second term is particularly interesting and will now be shown to relate to the geometric stiffness matrix usually associated with large deflection problems of structural analysis.

Let the equation $N^TF = P$ refer to the node equilibrium equations written in some specified geometry. A small variation from this equilibrium position can then be written as

$$d(N)F + N(dF) = dP$$

For buckling problems, a system is said to be unstable if a small perturbation of load can produce a large response (displacement) and it is customary to examine the variation indicated with respect to the change in node coordinates (the node displacement). It is convenient to write

$$d(N^T)\, F = (\nabla N^T)F \,.\, \delta$$

and to identify the geometric stiffness matrix K_G as

$$K_G = (\nabla N^T)F$$

It follows that

$$\partial t/\partial x_i = (K^*_G\, \delta)_i$$

Here the asterisk is used to indicate that K_G has the dimension of the set S and not the dimension of the Lagrange multiplier matrix δ as it usually does.

For the most simple truss problem, t(F) has the form

$$t(F) = \sum_i |F_i| L_i$$

and it follows that since

$$N_{ij} = \partial L_i / \partial x_j$$

then for the truss problem the optimality condition is

$$N^*|F| = K^*_G \delta$$

where the asterisk again indicates the reduced dimension of the problem. Generally, a two-step solution procedure can be used in which the forces are fixed and the geometry is allowed to vary (this is similar to the statically determinate case) and then the coordinates are fixed and the forces are allowed to vary.

5.3.1 Frame Problems

For the most simple case of plane frames discussed above, the objective function becomes

$$t(F) = \sum_i L_i \max\left\{\left|m_i^+\right|, \left|m_i^-\right|\right\}$$

and the optimality equation becomes

$$N_{truss}^{*T}\varphi = K_G^*\delta$$

where

$$\varphi = \max\{| m_i^+ |, | m_i^- |\}$$

Consider the example shown in Fig. 5.8. The geometry and loads indicated as stage 1 are prescribed initially and the algorithm then computes an optimum moment distribution, which is also indicated. The algorithm next tries to compute an improved geometry and its associated distribution of moments which are indicated as stage 2. Note that the algorithm was allowed to run through four stages and that the loaded points were constrained to move vertically.

As the optimization continues, it becomes typically more and more difficult to achieve an improved design and the algorithm was stopped in this case after four stages. For the example shown in Fig. 5.8 there unfortunately exists an optimal (funicular) design which is a very flat arch (almost a straight line) which has no bending moment at all and therefore a zero value of the objective function. There also exists a funicular design (again zero objective function) for any fixed value of the rise of the arch. In general, a more practical objective function, one which reflected the effect of axial load, would remove this degeneracy. But under the conditions given, the algorithm used seems to perform remarkably well.

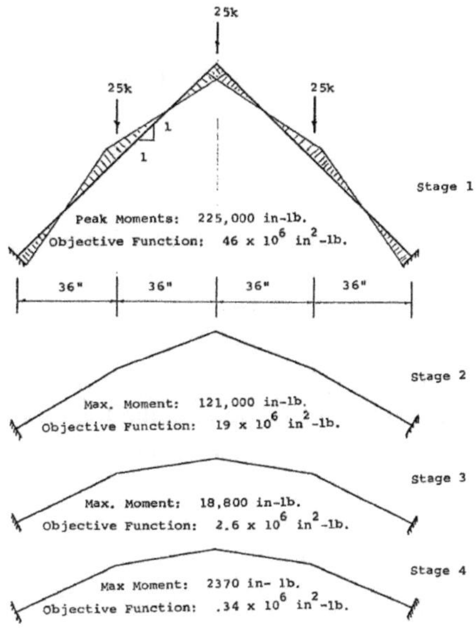

Fig. 5.9. An Arch Example

The final frame example of this section is shown in Fig. 5.9. It is more pathological and thus less interesting, but tends to underline the above comments. At the final shape indicated in this figure, the algorithm was stopped, anticipating that the member was trying to straighten itself out and the crooked leg was replaced by a single straight piece.

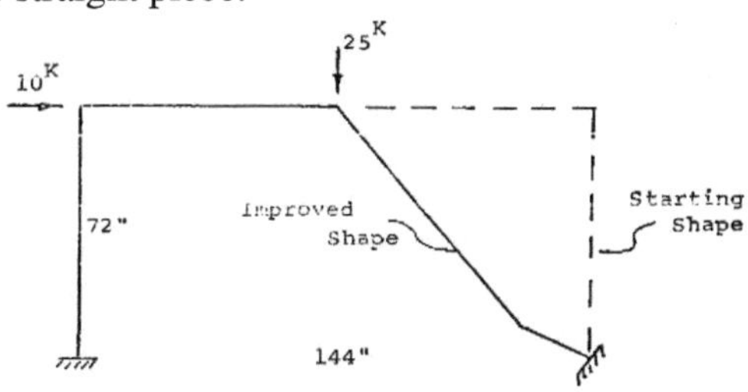

Fig. 5.10. A Rigid Frame

5.3.2 Code Design for a Truss Problem

Figure 5.10 shows a truss design problem that uses the objective function

$$\textit{minimize} \quad \sum |F_i|\, L_i$$

Obviously, this objective function does not reflect the penalty of code design for compression members over tension members. As a result of this omission, the design ends up with some rather long compression members. When this design problem has been scaled up (Fig. 5.11) in terms of both loads and dimensions so that code design can be applied, a different shape appears and the long compression members disappear. How this can be done is the topic of this section. For more details than presented here the reader is referred to Spillers and Kountouris (1980).

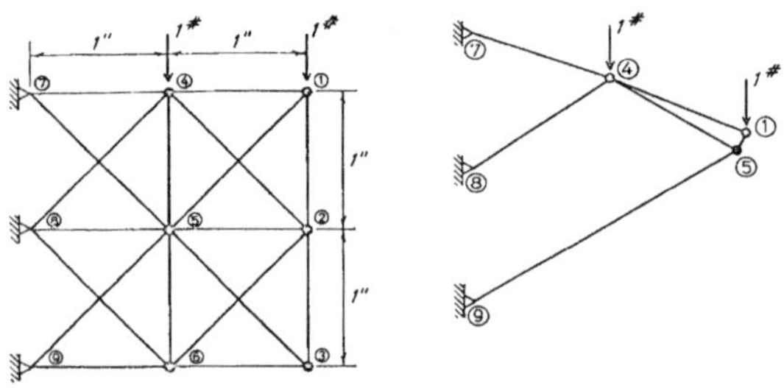

Fig. 5.11. Optimal Geometry for a Truss Design

There are several aspects of code design that make it difficult. Certainly one aspect is the fact that members must be selected from commercially available sections such as those in the AISC Manual as has been done in this example. The approach taken was to work with member forces as the engineer does manually. In this case two-angle struts 3/8″ back to back were considered. It is then possible to curve-fit a relationship between the radius of gyration r and the allowable compressive member force f for some specific member length L as

$$r = a + bf + cf^2$$

(See the table of these coefficients that is given below.) This in turn allows an expression to be written for the allowable compressive stress σ following the code to be written as

$$\sigma = \sigma(f, L) \qquad \sigma \geq 0$$

The objective function for this code design can then be written as

$$t = \sum_i \frac{\sigma_i^a}{\sigma} |F_i| \Delta_i^a$$

In this equation σ_i^a and Δ_i^a are simply convenient constants. In the paper cited above, a Lagrangian approach was taken to the solution of the optimization problem but we leave the details or even the approach to the solution to the reader.

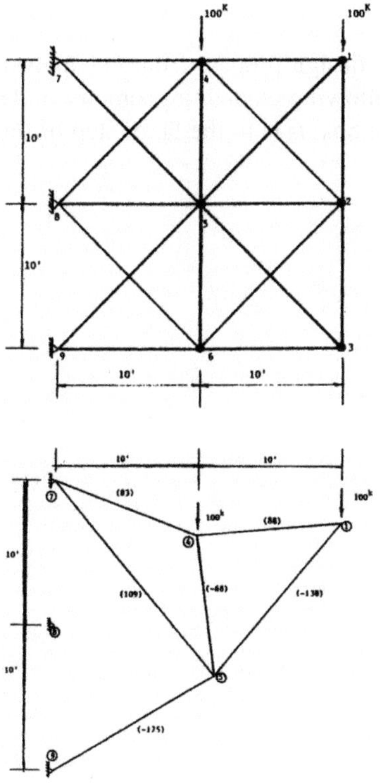

Fig. 5.12. A Truss Designed Using the AISC Code

	Coefficients*		
Length, in feet	*a*	*b*	*c*
5	0.619713	0.00702525	−0.00000691048
10	0.676079	0.00886255	−0.0000104503
15	0.752197	0.019025	−0.0000156750
20	0.816942	0.0132710	−0.0000238004

*$r = a + bf + cf^2$.

5.4 Generating New Designs Automatically

There is an almost obvious sequence within the design process from (1) analysis to (2) the optimization of a given design to (3) allowing changing geometry in design to (4) automatically creating new configurations. It is to the latter step of this sequence that this section is devoted.

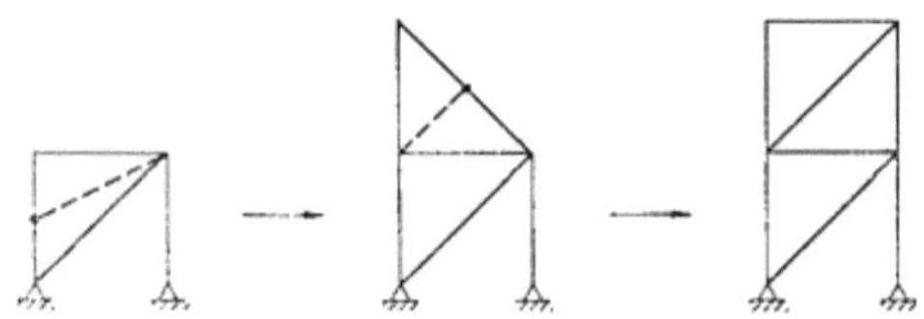

Fig. 5.13. Building with Triangles

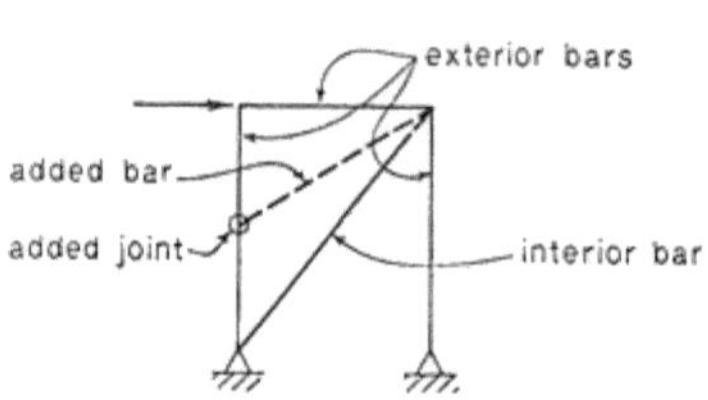

Fig. 5.14. Adding a Member

Figures 5.12 and 5.13 set the stage for this process. The idea is that there are many rules that can be developed for constructing new designs. The one shown here is referred to as building designs from triangles (Fig. 5.14). The idea is that you can start with a simple design and create automatically something that is *new*. In this case the design starts with the relatively simple case on the left of Fig. 5.15 and using the rule of Fig. 5.13 and geometric optimization ends up with the design on the right of Fig. 5.15 which is quite different from the starting design that has reduced weight and has the appearance of a Mitchell truss.

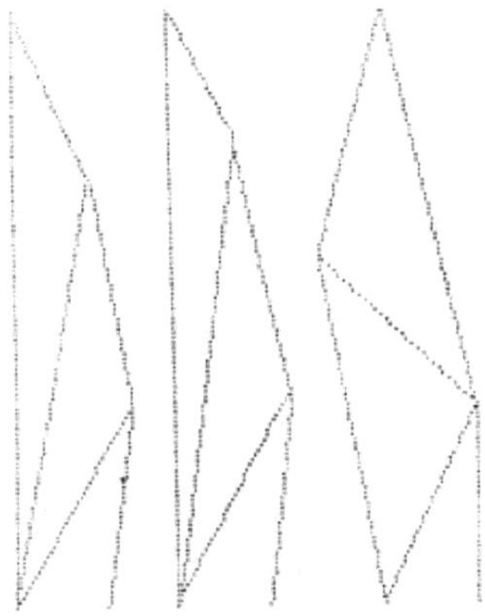

Fig. 5.15. Three Structures Whose Weight is Compared

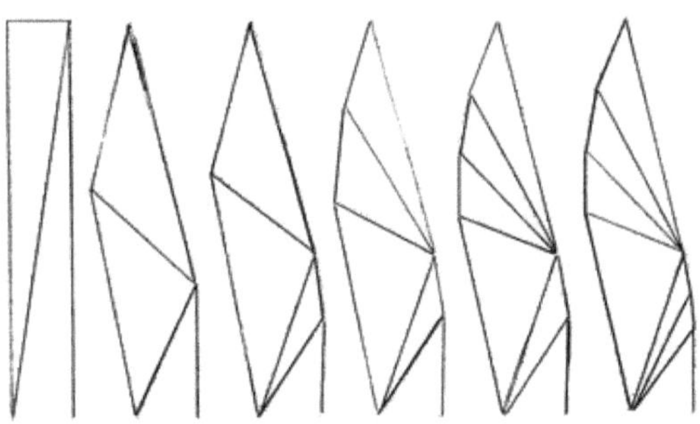

Fig. 5.16. Sequence of Examples with Decreasing Weight

Automating the configuration or topology of a structural design is an active area of research (cf., for example, Bendsoe and Soares, 1992). There are many ways to approach such a problem. Clearly, if the only question is one of connectivity then it is a *graph problem*. If the discussion is to include information about the structure itself then matters are more complicated and topology is only one facet of the problem. And there are technologies from circuit theory (Harrison, 1965) that can be useful in spite of the fact that electrical engineers do not have geometry to worry about.

In the remainder of this section some of the alternative methods for dealing with automating a structural configuration are discussed.

5.4.1 Algebraic Methods

In this section so-called algebraic methods are described. They are motivated by the idea that since Taylor series are basic to analysis and the representation of *functions*, why not use some variation of Taylor series to describe structural connectivity?

This discussion starts with the node–node matrix A_{ij} from graph theory that is subsequently mapped into a column matrix α:

$$A_{ij} = \begin{Bmatrix} 0 \ \textit{if nodes i and j are not connected by a branch} \\ 1 \ \textit{if nodes i and j are connected by a branch} \end{Bmatrix}$$

A_{ij} can be represented as a triangular array I with

$$I = \begin{matrix} A_{12} & A_{13} & A_{14} & \cdots & A_{1n} \\ & A_{23} & A_{24} & \cdots & A_{2n} \\ & & A_{34} & \cdots & A_{3n} \\ & & & & \vdots \\ & & & & A_{n-1,n} \end{matrix}$$

and I mapped into a column matrix α:

$$\alpha = \begin{bmatrix} A_{12} \\ A_{13} \\ A_{23} \\ A_{14} \\ A_{24} \\ A_{34} \\ \vdots \\ A_{1n} \\ A_{2n} \\ \vdots \\ A_{n-1,n} \end{bmatrix}$$

In order to make the machinery of vector spaces available, it is convenient to define multiplication and addition modulo 2 as follows:

y \ x	0	1
0	0	1
1	1	0

x + y

y \ x	0	1
0	0	0
1	0	1

xy

This allows graphs as binary matrices to be manipulated in an ordinary manner algebraically.

In linear algebra a nilpotent matrix T is an n × n array with the property that $T^{n-1} \neq 0 \quad while \quad T^n = 0$. The nilpotent matrix has the additional property that if α is an n×1 matrix such that $T^{n-1}\alpha \neq 0$, the matrices

$$\alpha, T\alpha, T^2\alpha, \ldots, T^{n-1}\alpha$$

are linearly independent and form a basis. This suggests that, in certain cases, an example α together with a nilpotent matrix T may be used to represent an arbitrary graph β as

$$\beta = \sum_{i=1}^{n} a_i T^{i-1}\alpha$$

The coefficients a_i of this expansion may be determined recursively as

$$T^{n-1}\beta = a_1 T^{n-1}\alpha$$
$$T^{n-2}\beta = a_1 T^{n-2}\alpha + a_2 T^{n-1}\alpha$$
$$T^{n-3}\beta = a_1 T^{n-3}\alpha + a_2 T^{n-2}\alpha + a_3 T^{n-1}\alpha$$

...

A more interesting question is concerned with the problem of determining the nilpotent matrix (operator) T which can be used to generate new graphs given a set of examples α_1 , α_2 , α_3 , ... It is convenient to start this discussion by observing that the commonly cited nilpotent matrix

$$T = \begin{bmatrix} 0 & 0 & 0 & 0 & 0 \\ 1 & 0 & 0 & 0 & 0 \\ 0 & 1 & 0 & 0 & 0 \\ 0 & 0 & 1 & 0 & 0 \\ 0 & 0 & 0 & 1 & 0 \end{bmatrix}$$

or more generally

$$T_{ij} = \begin{cases} 1 \; if \; i = j+1 \\ 0 \; otherwise \end{cases} \qquad (i, j = 1,2,\ldots,n)$$

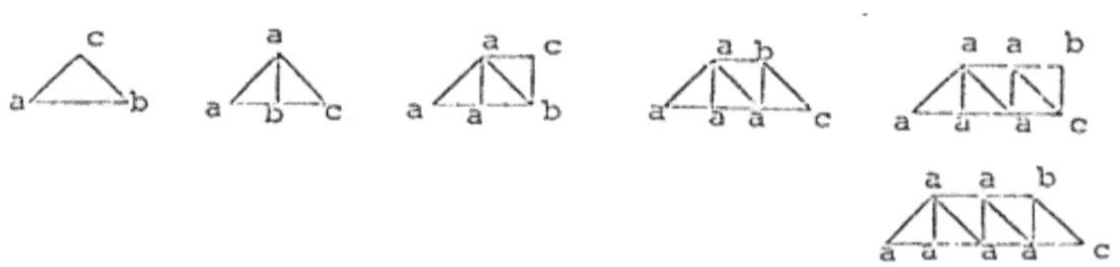

Fig. 5.17. Building with Triangles

This matrix tends to shift down one element, the nonzero elements of α in the product Tα. Note that the progression of graphs in Fig. 5.16 can be described as a shift, but a slightly more complex one since

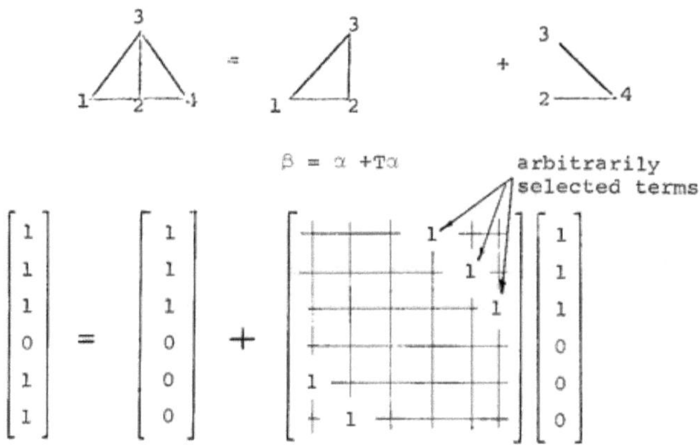

Generalizing upon these remarks several comments may now be made:

1. Where new graphs are constructed from old by simply adding branches and nodes and where the number of branches added is smaller than or equal to the number of original branches, this process can be carried out formally using a shifting operator on the matrix of the graph.

2. This shift can be described using nilpotent matrices.

3. The examples may not completely define the nilpotent matrix.

Under the conditions of Comment 1, the representation of the new graph differs in general from the representation of the old graph by the addition of new terms to the lower portion of the matrix of the old graph. In fact, the notation was constructed so that this would be the case. The second comment is based upon the fact that if R is a permutation matrix, the nilpotent matrix T remains nilpotent under the transformation R^TTR so that the more general shifts required above are also representable through nilpotent matrices. Finally, Comment 3 refers to the fact that the matrix T in the above example is not a nilpotent matrix but can be made into 1 through the addition of terms to which the example is indifferent. In this sense the matrix may be called "incomplete" and it is in some respects reasonable that a single new example does not define the operator completely. In fact, when another piece is added to the graph in this fashion, the size of the arrays increases and the problem of the incomplete operators persists.

The case shown in Fig. 5.17 is a little more interesting and will be discussed now. Note first that with proper node numbering, the case shown in Fig. 5.16 can be described by adding, at each step, the two bars and the new node to the last two nodes added to the graph. In these terms, the case shown in Fig. 5.18 can be described in the same way,

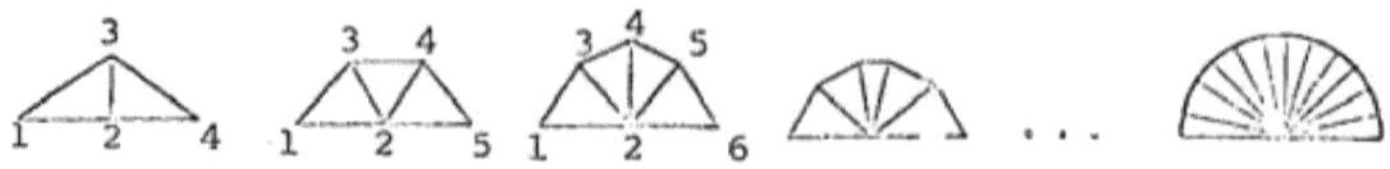

Fig. 5.18. Convergence to Mitchell's Result

but it also requires an interchange at each step of the numbers of the next two last two nodes. In any case, a possible matrix representation follows the comments made above.

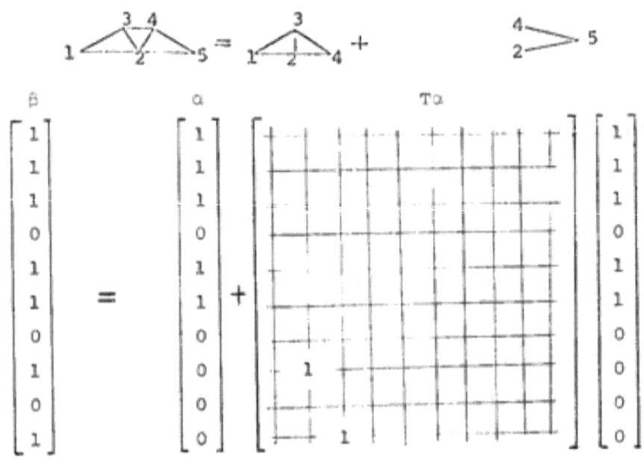

Rather than simply using matrix algebra to represent the composition of designs, it is more interesting to develop matrix operators that can be used subsequently with new examples. It was possible in the case shown in Fig. 5.17 to arbitrarily select terms so that the progression of designs could be represented by the equation a + Ta + T^2a +··· where T is a nilpotent matrix. In view of the form of the example associated with Fig. 5.17 that is no longer the case. As a final example it will be shown how this problem can be solved by starting with a smaller, equivalent, example and then how the derived transformation can be used with a new example. If the previous case is written as

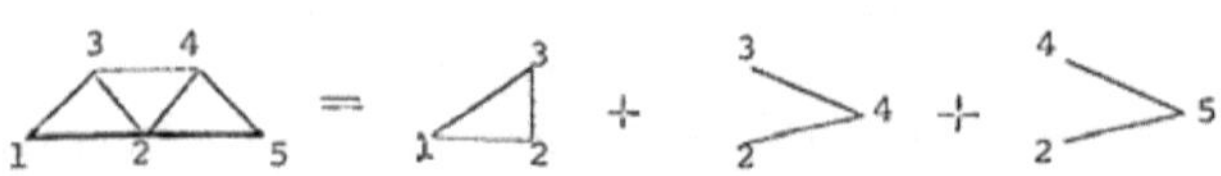

$$\beta = \alpha + T\alpha + T^2\alpha$$

the nilpotent matrix T can be constructed to be

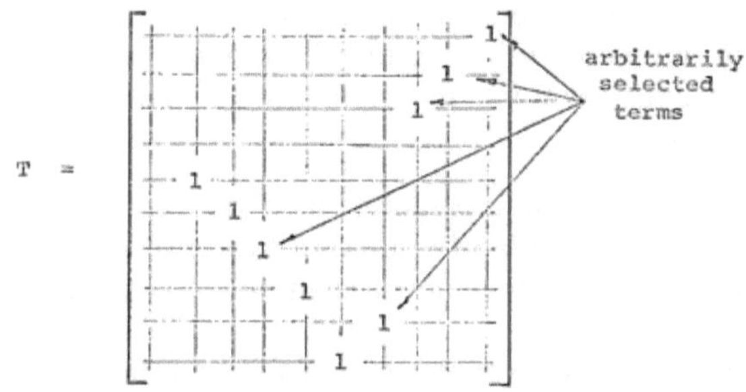

Since the matrix T is a permutation of the classic nilpotent matrix cited above, it has the desired properties and can be subsequently used with other examples to create new designs. This is illustrated in Fig. 5.19.

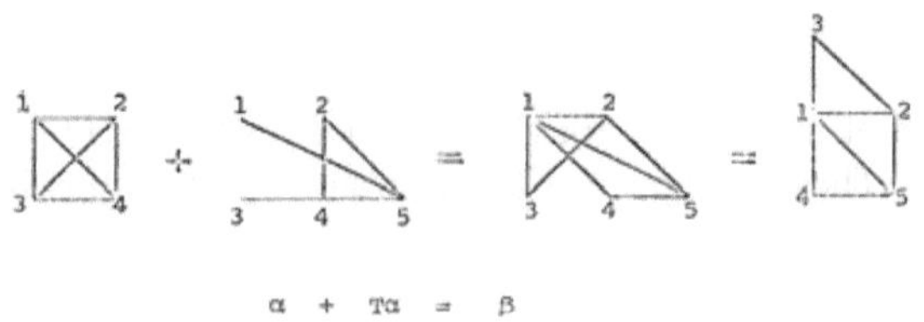

Fig. 5.19. Using the Matrix T

5.4.2 Linguistic Methods

There has been a considerable amount of work recently directed toward the application of linguistic methods to picture processing and pattern recognition. In this section both the application of these techniques and their potential in structural design will be discussed.

Quite in general, subsets of pattern recognition and picture processing are concerned with pictures that can be represented by graphs and in that sense the development of web (graph) grammars is natural. Unfortunately, the question of the type of application arises here. In particular, in picture processing there is a concern for the semantics of a picture, which is not immediately shared by structural design.

Central to linguistic methods is the idea of a grammar (in the case of graphs a web grammar) from which a language (a set of graphs) can be generated. This section will follow the work of Pfaltz and Rosenfeld (1969) and define a web grammar G to be a triple of the form (V,I,R), where V is a vocabulary, I is a set of "initial" webs, and R is a set of rewriting rules. Without redeveloping the terminology of graph theory:

1. The vocabulary V is a finite set of elements which is to be used, in conjunction with the "initial webs" to generate the language. It is common practice to divide the set V into the "non-terminal" vocabulary V_N and the "terminal" vocabulary V_T.

2 The set of initial webs, I, defines initial configurations from which new graphs are to be generated. (In an adaptive design system they would represent examples from which new designs are to be generated.)

3 Each rewriting rule of R consists of a triple (α, β, E), where α and β are webs and E is an embedding which defines how web β is to replace web α in the web ω – α where ω is the host web. The rewriting rules will use the notation (a,b) | (c,d) to indicate that branch (c,d) in the rewritten web replaces branch (a,b) in the host web. The symbol p is used as a generic node label: the symbol := is used to indicate that a subgraph "is to be replaced by" another subgraph.

Figure 5.16 shows a progression of examples through which a designer might proceed in a truss design problem. These designs can be generated by the following grammar:

Grammar 1.

V = {A,B,C,a,b,c} where the capital and the lower case letters denote V_N and V_T respectively.

I =

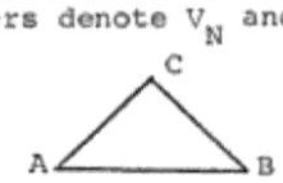

Rewriting Rules

1)

E = {(A,B)|(C,B) , (B,a)|(B,A) , (A,a)|(C,A), (a,p)|(A,p) , (A,p)|(C,p) , (B,p)|(B,p)}

2)

E = {(B,A)|(C,B) , (A,a)|(B,A) , (B,a)|(C,A), (a,p)|(A,p) , (A,p)|(A,p) , (B,p)|(C,p)}

3) A : = a E = {(a,p)|(A,p)}

4) B : = b E = {(b,p)|(B,p)}

5) C : = c E = {(c,p)|(C,p)}

With this introduction to design automation, we leave the reader to pursue this topic with books such as Knight (1994) and Stiny (2006).

5.5 Problems

1. Verify the dual of the two loading condition problem by first converting the primal to a standard linear programming problem and then converting the standard linear programming dual to the dual indicated.
2. Verify the dual of the three loading condition problem as suggested in problem 1.
3. Venkayya (1971) discusses the three-bar truss of Fig. 5.19. In this case the inclined bars slope 1:1, E =1.× 10^7 psi, the allowable stress in the inclined bars is 5000 psi, and in the vertical bar, 20,000 psi.

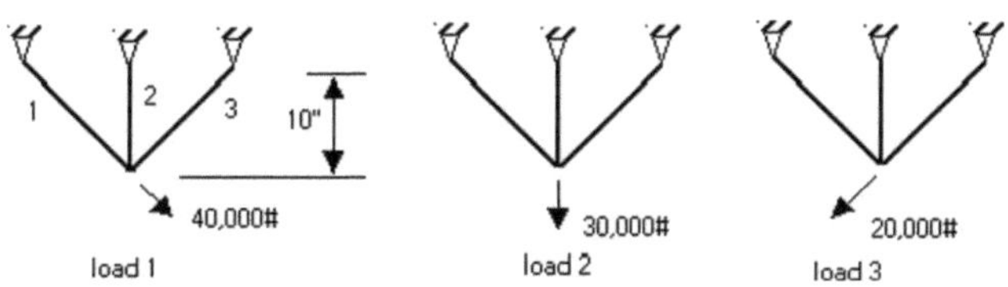

Fig. 5.19. Three Loading Conditions

Venkayya cites the optimal volume of material in this case to be 159.86 in^3. Following the two loading condition problem above the optimal design (Program 6.3) was determined to be A_1=0.70, A_2=2.13, and A_3=2.76 for a volume of 159.8 in^3. And there is a statically determinate design (member 1 omitted) with a volume of 141.2 in^3.
Discuss the fact that the behavior of this problem is not continuous.

4. Solve problem 3 above using the truss program from Chapter 3.
5. Find the optimal realizable solution for the two loading condition problem of Section 5.1.1. Discuss the effect of the use of prestress in your solution.
6. Verify the solution of the truss in Fig. 5.4 using the truss program from Chapter 3.
7. Outline a frame program for geometric optimization using the AISC code.

6 Beams and Plates: The Work of Rozvany

This chapter discusses some problems of the design of optimal beams and plates that depend largely on the work of G. I. N. Rozvany (1976). (See also Rozvany et al., 1995). He makes particularly clever use of Lagrange multipliers as discontinuous functions.

6.1 Introduction

Clearly, flexural systems are basic structural elements. The most comprehensive text dealing with flexural systems is Rozvany's book cited above. That book is the focus of this chapter.

We will first of all discuss a sequence of beam problems that center on the problem mentioned earlier in this text of finding the bending moment, m, in a beam to

minimize $\int |m|\, dx$ subject to $m'' = -w$ (beam equilibrium)

with w the lateral load on the beam. This formulation assumes, for example, that the area of the beam cross section is proportional to the absolute value of the bending moment. The solution discussed earlier involved forming the Lagrangian

$$L = \int f(m, m'')dx = \int \{|m| + \lambda(m'' + w)\}dx$$

and then the Euler–Lagrange equation implies that

$$\frac{\partial f}{\partial m} - \frac{\partial}{\partial x}\left(\frac{\partial f}{\partial m'}\right) + \frac{\partial^2}{\partial x^2}\left(\frac{\partial f}{\partial m''}\right) = 0 \qquad \Rightarrow \qquad \operatorname{sgn}(m) = -\lambda''$$

It is worthwhile to note here that the integration by parts required by the Euler–Lagrange equation implies that both λ and λ' are zero at the ends of the integration interval.

It is common to regard the Lagrange multiplier λ to be a displacement, say y, and then λ'' becomes the curvature y''. The Euler–Lagrange equation then implies a displacement field in which the curvature has the opposite of the sign of the moment and has a magnitude of 1. In the spirit of virtual work the equilibrium equations can be multiplied by the displacement y and integrated over the span to give

W.R. Spillers, K.M. MacBain, *Structural Optimization*,
DOI 10.1007/978-0-387-95865-1_6, © Springer Science+Business Media, LLC 2009

$$\int m y'' \, dx = -\int w y \, dx$$

(integrating by parts again). Finally using the optimality condition, it follows that

$$\int |m| dx \geq \int (-m) y'' \, dx = \int w y \, dx$$

in which equality holds for an optimal design. It follows that either the moment diagram or the displacement field can be used to compute the value of the objective function for an optimal design.

6.1.1 Examples

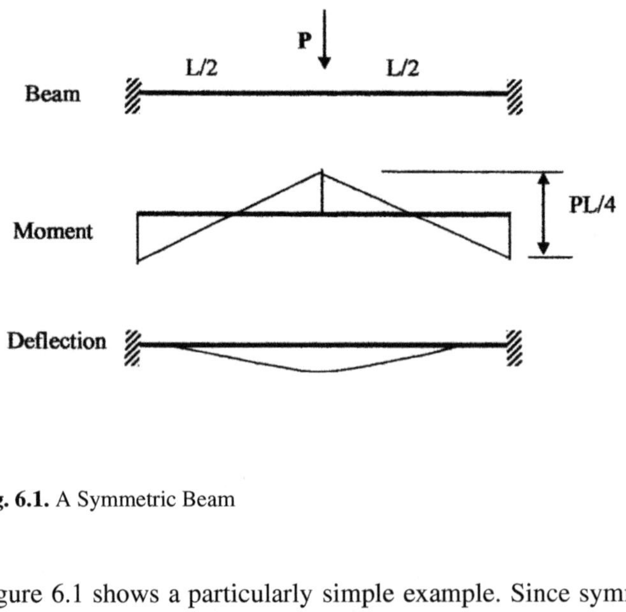

Fig. 6.1. A Symmetric Beam

Figure 6.1 shows a particularly simple example. Since symmetry implies that the displacement field has inflection points at the ¼ points, the optimal moment diagram then has values of $\pm$ PL/8 and the objective function is equal to ½ × PL/8 × L/4 × 4 = $PL^2/16$. Using the displacement field to compute the objective function:

$y''=1 \;\Rightarrow\; y=x^2/2 \;\Rightarrow\; y_{center} = 2 \times (L/4)^2 \times 1/2 \;\Rightarrow\;$ work done = $P\,(L^2/16)$

as indicated above.

The case of a uniformly distributed load is somewhat more complicated. Note first of all that equilibrium implies that the optimal moment diagram must be the simply supported moment diagram upon which the effect of the end moments has been superimposed (Fig. 6.2).

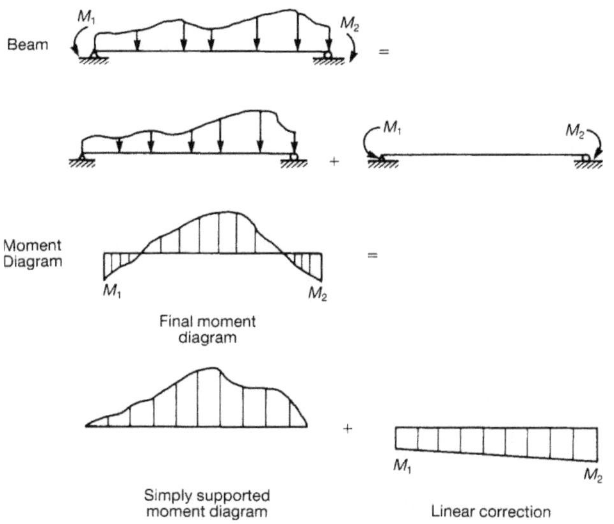

Fig. 6.2. Superposition of Moment Diagrams

The moment diagram for the uniformly loaded beam is indicated in Fig. 6.3. Since it is parabolic and using the fact that the moment must be zero at the quarter point, the end moments must be $wL^2/8 \times (1-1/4) = 3/32\ w\ L^2$ (see Fig. 6.4).

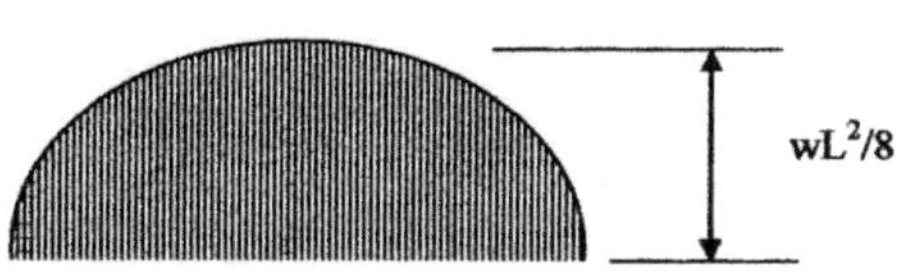

Fig. 6.3. Moment Diagram for a Uniformly Loaded Simply Supported Beam

The optimal moment diagram can then be formed by drawing a horizontal line through the simply supported moment diagram shown above. The line must be drawn so that the moment at the ¼ point is zero. Since the moment diagram in this case is parabolic, the maximum positive moment is ¼ × $wL^2/8$.

Fig. 6.4. Optimal Moment Diagram

Again, the value of the objective function can be computed either from the moment diagram or from the deflection curve. For the case of the former, when x is measured from the center of the span

$$M = \frac{1}{4}\frac{wL^2}{8} - \frac{wx^2}{2} \quad \Rightarrow \quad 2\int_0^{L/2} |M| dx = wL^3/32$$

For the case of the latter and using the above example

$$\int wy dx = 2w\left\{ \int_0^{L/4} (x^2/2)dx + \int_{L/4}^{L/2} (L^2/16 - x^2/2)dx \right\} = wL^3/32$$

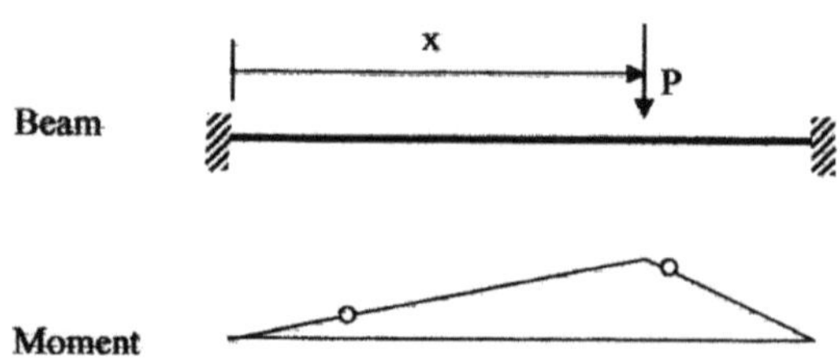

Fig. 6.5. Load at Some Point x

In fact, this process of constructing an optimal moment diagram is rather general. Consider the case of a beam with a concentrated load at some point x as indicated in Fig. 6.5. The optimal moment diagram can be obtained again by drawing a line between the moments at the ¼ points of the simply supported moment diagram.

This construction breaks down when the load, for example, moves to the right of the ¼ point. But in that case the optimal design becomes a cantilever.

6.1.2 Reaction Costs

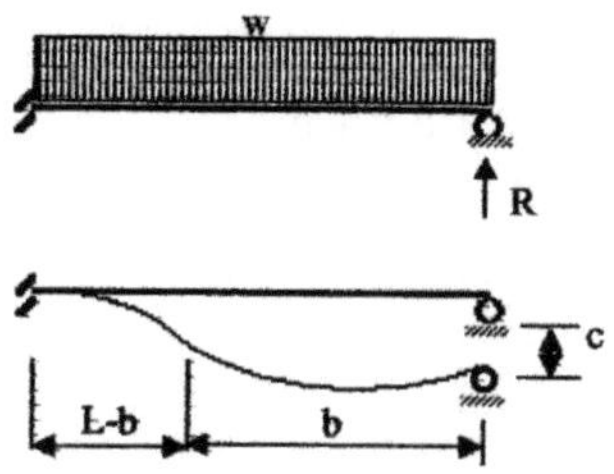

Fig. 6.6. A Propped Cantilever

In this case it is desired to include the cost of the reaction R in Fig. 6.6 in addition to the cost of the beam itself. The optimization problem to be solved is then

$$\text{minimize} \quad \int |M| dx + cR \quad \text{subject to} \quad M'' = -w$$

Here R is the reaction at the right end of the beam and c is the cost of the reaction per unit of force. In order to reduce this problem to a form similar to the problem just discussed, two steps are needed. First, the reaction can be written in terms of the beam shear at the right end as R = – M′. Then, the reaction cost can be included in the objective function as

$$\text{minimize} \int_0^L \{ |M| - c\delta(x - (L - \varepsilon))M' \} dx \quad \text{subject to} \quad M'' = -w$$

Here δ is the Kronecker delta function and ε is a small number. The idea is, of course, to include the reaction costs in the integral. The Euler equation then becomes

$$\text{sgn}(m) + c[\,\delta(x-(L-\varepsilon))]' + y'' = 0$$

This equation implies that the curvature has a higher order discontinuity (δ′) but that when you integrate twice to obtain the displacement it has the step discontinuity indicated in Fig. 6.6 since the delta function is the derivative of the Heaviside step function.

To complete this example, it is only necessary to locate the inflection point (the point of zero moment) that is described by the term b in Fig. 6.6. That can be done as follows:

Left of inflection point	Right of inflection point
$y'' = 1$	$y'' = -1$
$y' = x$	$y' = -x + c_1$
$y = x^2/2$	$y = -x^2/2 + c_1 x + c_2$

The boundary conditions are

$$y_{left} = y_{right} \quad \text{at } x = L - b$$
$$y'_{left} = y'_{right} \quad \text{at } x = L - b$$
$$y_{right} = c \quad \text{at } x = L$$

It follows that

$$b = \sqrt{L^2/2 - c}$$

6.1.3 Optimal Segmentation

In the same spirit, Rozvany considers the problem of optimal segmentation. Here a uniformly loaded, simply supported beam is to be made up of two uniform segments. The question is where to locate the change in cross section which is described by the parameter α depicted in Fig. 6.7.

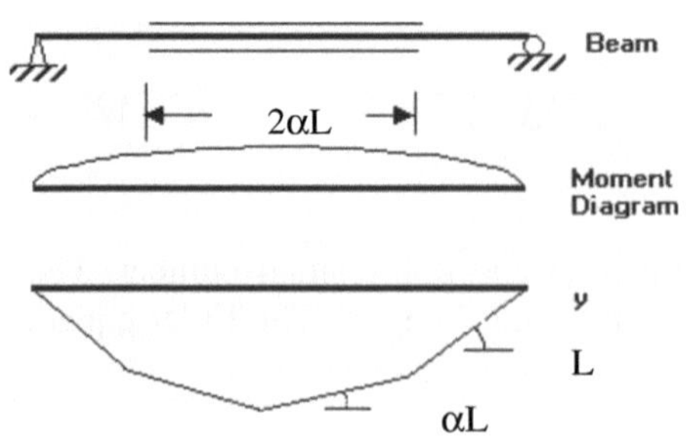

Fig. 6.7. Beam with a Cover Plate

In the spirit of the other problems of this chapter, an attempt is made here to find α to

minimize $m_0 \alpha L + m_1 (L - \alpha L)$ subject to equilibrium

Here m(x) is the given moment diagram, m_0 is the center moment, m_1 is the moment at the end of the cover plate, and the span length is 2L. The idea is to use the delta function to *turn on* the moment at the center and then the moment at the end of the cover plate. The optimization problem is then

minimize $=2 \int [(\delta(x)|m| \alpha L + \delta(x - \alpha L) |m| (L-\alpha L)] dx$ subject to equilibrium

with x measured from the center of the beam. The Euler equation then becomes in this case

$$\mathrm{sgn}(m)\, \delta(x)\, \alpha L + \mathrm{sgn}(m)\delta(x - \alpha L)\ (1-\alpha)L + y'' = 0$$

When this equation is integrated to obtain the slope, the results shown in Fig. 6.7 can be seen. The optimal α can be determined directly from the objective function. For a statically determinate beam with a uniform load α can be computed directly. In this case the objective function can be written as

$$\varphi = \frac{wL^2}{2}\alpha L + \left(\frac{wL^2}{2} - \frac{w}{2}(\alpha L)^2 \right)(1-\alpha)L$$

Setting $\partial \varphi / \partial \alpha = 0$ in this case gives the optimal $\alpha = ½$.

6.1.4 A More Realistic Beam Model

In order to use a more *realistic* beam model it is common to assume that the weight of a beam as proportional to $|m|^n$ rather than simply using $|m|$. (This point has been used in some of the practical design examples of this text.) In that case the beam design problem of this chapter becomes

minimize $\int |m|^n dx$ subject to $m'' = -w$ (equilibrium)

Proceeding as before, the Lagrangian is formed as

$$L = \int \{ |m|^n + \lambda (m'' + w) \} dx \quad \Rightarrow \quad n\, | m |^{n-1}\, \mathrm{sgn}(m) + \lambda'' = 0$$

For the case of n=2 and a fixed ended beam with a uniform load this problem can be solved directly. Using some results from the earlier problems,

$$\varphi = \int m^2 dx = 2\int_0^{L/2}\left(\frac{wL^2}{8} - \frac{wx^2}{2} - \overline{M}\right)^2 dx$$

Where $\overline{M}$ is the moment at the support. If we set $\partial\varphi / \partial\overline{M} = 0$ and then do the integration, it follows that $\overline{M} = \frac{wL^2}{12}$.

6.2 Design of Plates

In 1972, Rozvany and Adidam published Figs. 6.8 and 6.9. These figures relate to the optimal design of plates and have considerable practical significance for structural engineers as they design floor slabs. In the work that follows we will reconstruct the design they show for a simply supported, square plate under a uniform load p.

Formally, Rozvany and Adidam solve the plastic design problem from plate theory attempting to find the plate bending moments M_x, M_y, and M_{xy} that

$$\textit{minimize} \int_S \left(|M_1| + |M_2|\right) dS$$

subject to

$$\frac{\partial^2 M_x}{\partial x^2} + \frac{\partial^2 M_y}{\partial y^2} - 2\frac{\partial^2 M_{xy}}{\partial x\,\partial y} = -p \qquad (\textit{plate equilibrium})$$

Where M_1 and M_2 are the principal moments.

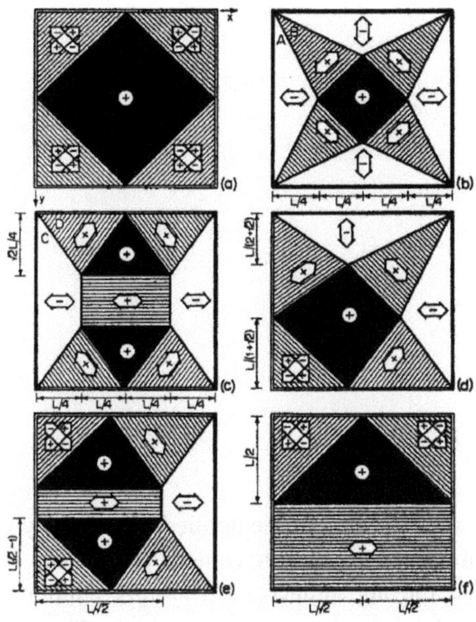

Fig. 6.8. Optimal Plate Layouts (Rozvany and Adidam, 1972)

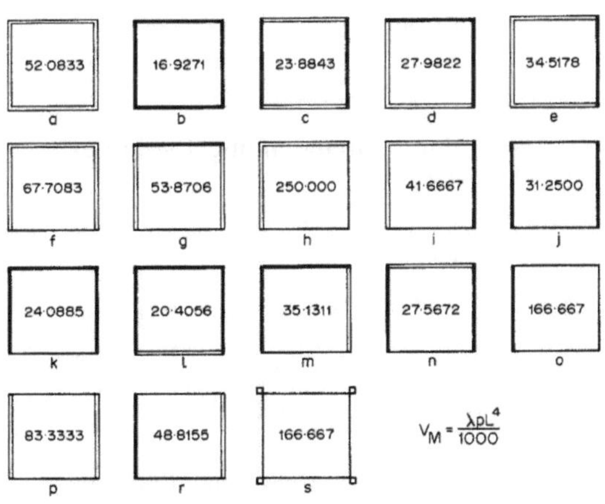

Fig. 6.9. Moment Volumes (Rozvany and Adidam, 1972)

In order to reconstruct the Rozvany/Adidam design it is only necessary to construct an appropriate *equilibrium* field since this is plastic design. How that is done is indicated in Fig. 6.10. The design, first of all, consists of a grid of

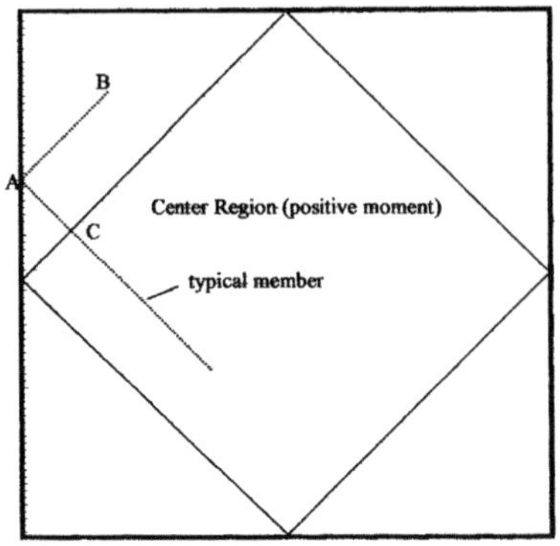

Fig. 6.10. Layout of Simply Supported Plate

beams crossed and inclined at 45° each sharing 50% of the applied load p. (Typical elements are indicated as ABC in the figure.) Over the center of this plate the beams have positive moment and have the appearance of a simply supported beam. At the edges, for example, the segment AC has negative bending while the segment AB has positive bending moment. At their intersection, point A, the moments combine to give a resultant moment perpendicular to the edge thus corresponding to a simply supported edge. The details of this design are now indicated.

Center Section:

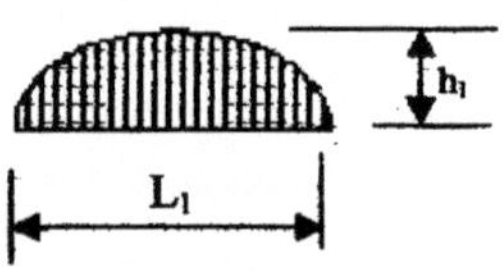

$h_1 = p(L/2^{1/2})^2/8$
$L_1 = L/2^{1/2}$

Contribution to the moment volume:
$pL^4/48$

Negative Moment Section:

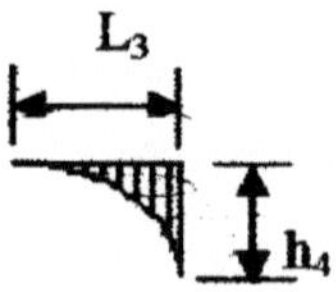

$L_3 = x$ (varies from 0 to $L/(2\ 2^{1/2})$)
$h_4 = pL/(4\ 2^{1/2})x + p\,x^2/4$

Contribution to the moment volume:
$0.03125\ pL^4$

(8 pieces)

Positive Corner Moment Section:

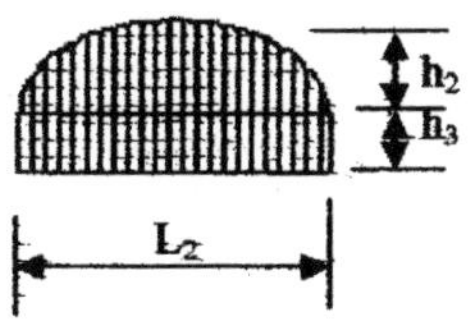

L_2 varies as x does above
h_2 is due to the distributed load
h_3 same as h_4 above

Contribution to the moment volume included in the negative moment section

The total moment volume is 0.052 p L^4 as shown in Fig. 6.9.

6.3 Problems

1. Reconstruct the design shown in this chapter for a square plate with fixed supports.
2. Try other loading conditions as optimal beam design problems.

7 Some Problems of Dynamic Structural Optimization

This chapter discusses some dynamic problems that have their roots in design for wind and earthquake loads. It includes both transient and steady-state vibrations.

7.1 Introduction

Dynamic problems of mechanics seem to fall into two categories: steady-state and transient vibrations. Steady-state problems (Cassis and Schmit, 1976) typically have the elements of the structure moving harmonically at the forcing frequency and occur, for example, when the wind blows uniformly on a building or the engines run at speed on a ship at sea. Transient vibrations occur, for example, when a building is subjected to an earthquake or a weight is dropped on a beam. The most simple example of the differences between steady-state and transient vibrations can be seen when a harmonic force is applied to a single degree of freedom system (Fig. 7.1).

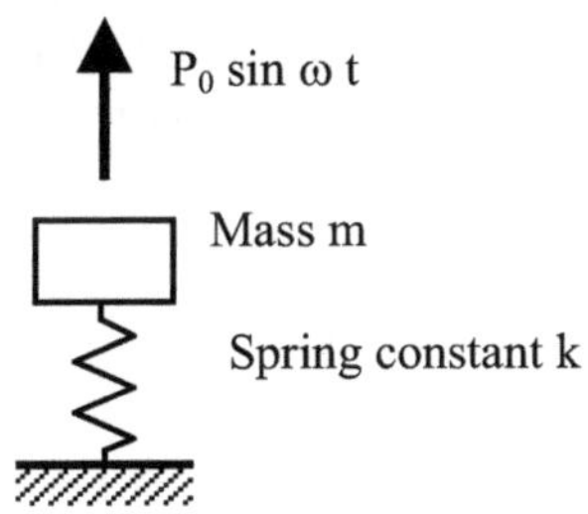

Fig. 7.1. Single Degree of Freedom System (SDOF)

The equation of motion is, in this case,

$$P_0 \sin \omega t - ky = m\ddot{y}$$

The solution has two parts: There is the so-called particular solution

$$y = constant\ x \sin \omega t \quad \Rightarrow \quad y_p = \frac{P_0 \sin \omega t}{k - m\omega^2}$$

and the so-called homogeneous solution

W.R. Spillers, K.M. MacBain, *Structural Optimization*,
DOI 10.1007/978-0-387-95865-1_7, © Springer Science+Business Media, LLC 2009

$$-ky = m\ddot{y} \quad \Rightarrow \quad y_h = -\frac{P_0 \omega \sin \alpha t}{(k - \omega^2 m)\alpha} \quad with \quad \alpha^2 = k/m$$

that uses the initial conditions

$$y = \dot{y} = 0 \quad at \quad t = 0$$

The general solution is, of course, $y_p + y_h$. The steady-state solution is y_p, which has the mass moving harmonically at the forcing frequency ω. It is used with comments such as damping will eventually eliminate the homogeneous solution. But doing so does not allow you to satisfy the initial conditions of the problem. In the work that follows, both types of solutions will be considered.

7.2 Optimization for Transient Vibrations

This section describes some problems of the optimal shape of a cantilever beam subjected to transient loading. It represents a basic problem of beam design, it has some application to problems of the design of tall buildings, and beyond that, it is of some general interest since there is (Kang et al., 2006) little available work with regard to optimization for problems of transient vibration. Certainly, one of the reasons for this absence of optimization studies of transient dynamic systems is the fact that there is no simple mathematical description of the *maximum* response of a vibrating structure.

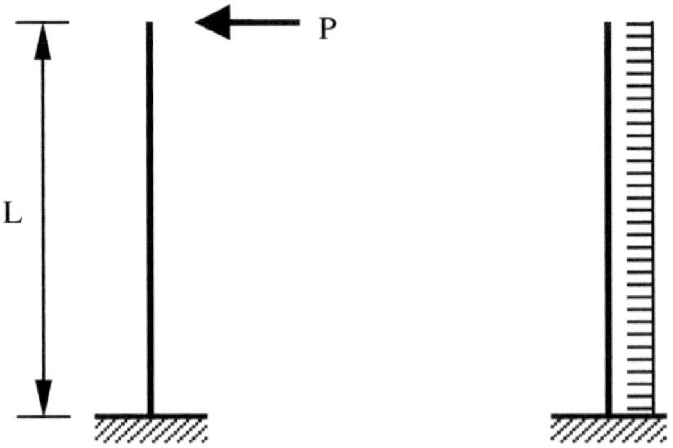

Fig. 7.2. Loaded Cantilever Beam

Figure 7.2 shows a cantilever beam subjected to two different loading conditions. The question is, when the amount of stiffness (material) is limited, where should it be placed to minimize the tip displacement? The solution to this problem is simple in the case of static loading. Dynamic loading will be seen to be more difficult to deal with.

The beam equations to be used in this section are the dynamic, flexural equations:

$$w + V' = \rho\ddot{y} \qquad M' = V \qquad M = -EIy''$$

(motion) (rotational equilibrium) (constitutive equation)

with

w – applied load
V – shear
ρ – mass density
M – bending moment
E – Young's modulus
I – moment of inertia
y – beam displacement

(The prime is used here to indicate differentiation with respect to the spatial variable x and the dot is used to indicate a time derivative.)

In this case, it is convenient to introduce non-dimensional variables indicated by the asterisk (that will subsequently be omitted):

$x^* = x / L$
$y^* = y / L$
$M^* = ML / (EI)_0$
$V^* = VL^2 / (EI)_0$
$w^* = w\, L^3 / (EI)_0$
$t^* = t\, [\rho\, L^4 / (EI)_0]^{1/2}$

It is also convenient to write EI as $(EI)_0\ \varphi(x)$ where $(EI)_0$ is a constant reference value and $\varphi(x)$ is a *shape function.* In this case the beam equations transform as

$$w - (EIy'')'' = \rho\ddot{y} \qquad \Rightarrow \qquad w - (\varphi y'')'' = \ddot{y}$$

Note that in the non-dimensional form, the constants have disappeared leaving only the function φ with x ranging over 0–1. The optimization problem can now be stated as find φ to minimize the tip displacement while keeping the *stiffness volume* constant.

7.2.1 The Static Case

The static case is relatively easy to deal with. Using the non-dimensional variables, the tip displacement for a *uniform* cantilever with a unit concentrated load is 1/3. The tip displacement for a uniform beam with a unit-distributed load is 0.125. For optimal designs the problem can be written as find φ to

$$\textit{minimize} \int_0^1 \frac{MM^V}{\varphi} dx \qquad \textit{subject to} \qquad \int_0^1 \varphi\, dx = 1$$

(That is, the stiffness can be moved around the beam but the *total stiffness* is to remain the same.) Here the displacement is computed using the virtual work method with M^V the virtual moment. This problem can be solved using the Lagrange multiplier method as

$$L = \int_0^1 \left[\frac{MM^V}{\varphi} + \lambda\varphi \right] dx \qquad \textit{and} \qquad \frac{\partial L}{\partial \varphi} = 0 \qquad \Rightarrow \qquad -\frac{MM^V}{\varphi^2} + \lambda = 0$$

Clearly, the stiffness φ is determined by the product of the moments and the Lagrange multiplier λ is determined by the integral constraint:

$$\varphi = \sqrt{MM^V / \lambda} \qquad \textit{and} \qquad \sqrt{\lambda} = \int_0^1 \sqrt{MM^V}\, dx$$

For the two cases of Fig. 7.2 (with x measured from the base of the cantilever),

Concentrated load: $M = M^V = -(1-x)$, $\sqrt{\lambda} = 1/2 \Rightarrow \varphi = 2(1-x)$

Distributed load: $M = -(1-x)2/2$, $M^V = -(1-x)$, $\sqrt{\lambda} = \frac{\sqrt{2}}{5} \Rightarrow \varphi = \frac{5}{2}(1-x)^{3/2}$

These stiffnesses are shown in Fig. 7.3.

7.2.2 The Dynamic Case

In this case, the loads are applied as step functions in time, that is

w (static case) → w H(t) (dynamic case)

where H(t) is the Heaviside step function. Putting aside for a moment the optimization problem, there is the beam equation

$$w - (\varphi y'')'' = \ddot{y}$$

to be solved. This is a parabolic partial differential equation. Since this equation is to be solved many times it seems reasonable to look for a *utility* to do so. Pennington and Barnes (1994) suggest that the *method of lines* is the most commonly used tool for problems of this type. (The method of lines has been championed by Schiesser (1991) and Wouwer et al. (2001))[1]. It was used in the calculations that follow and appears to us to give good results.

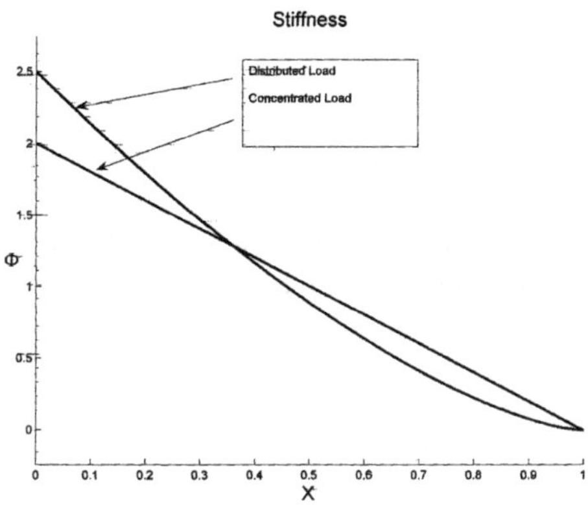

Fig. 7.3. Optimal Stiffness for the Static Case

Finally, the optimization problem can be approached. It involves several caveats. The absence of a simple mathematical description of the *maximum* is handled here by

[1] The software used in this section is a modification of the software of Professor Schiesser of Lehigh University (http://www.lehigh.edu/~wes1/wes1.html).

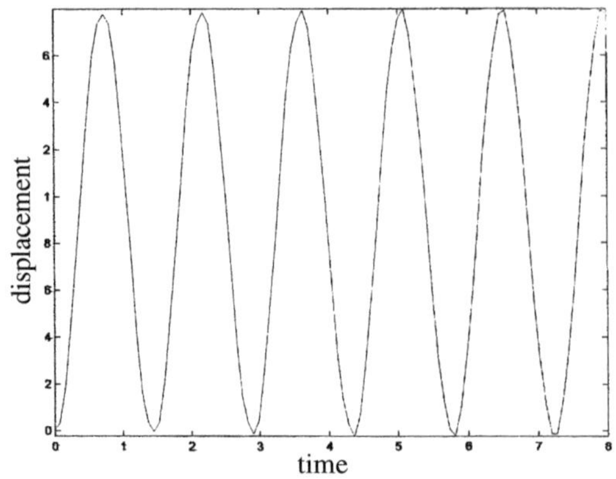

Fig. 7.4. Tip Displacement Versus Time for the Optimal Design

letting the solution run for an arbitrary period of time and then selecting the maximum calculated displacement. Since the solution is well behaved in this case (Fig. 7.4) that would appear to be a valid approach.

With regard to the variation of the function φ over the length of the beam, a kind of Taylor series approach involving two parameters, α and β, has been used. First a linear variation of stiffness with constant volume

$$\varphi = [1 - \alpha\,(x - ½)\,]$$

is used varying α systematically (see Fig. 7.5). We note that φ is a well-behaved function of α in this case.

In order to try a more complex (quadratic) shape, a new function

$$\varphi = [1 - \alpha\,(x - ½)\,] + \beta\,[\,-1/3 + (1 - x\,)^2]$$

is used. It can be seen that the *stiffness volume* is constant for any values of α and β. With this more complex shape function the search process is extended: For each

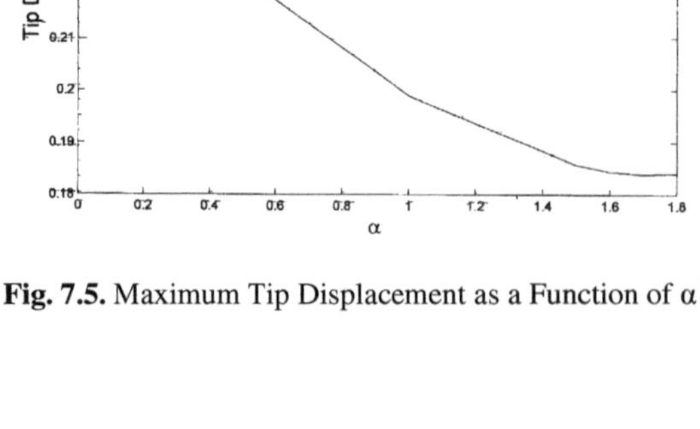

Fig. 7.5. Maximum Tip Displacement as a Function of α When β is Zero.

value of α, β is varied systematically. In both load cases the optimal values of α and β are approximately $\alpha = 1.5$ and $\beta = 0.4$. The optimal beam shape is shown in Fig. 7.6.

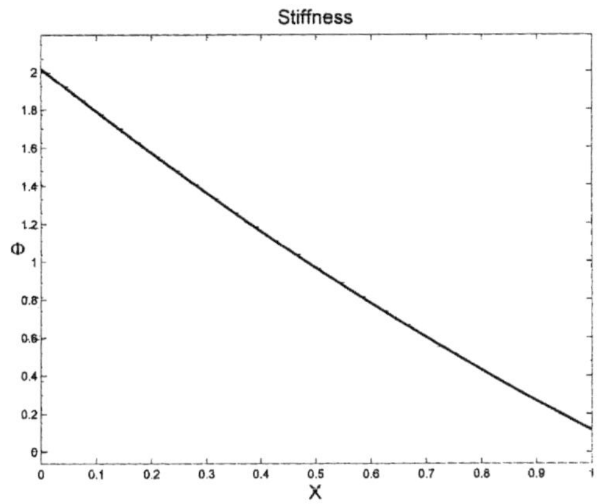

Fig. 7.6. Optimal Stiffness Distribution (Dynamic Case)

It seems clear that the optimal design for transient loads is a lot like the optimal design for static loads. The following table summarizes the calculations performed:

Tip Displacement

	Concentrated load	Distributed load
Static case		
Uniform beam	0.333	0.125
Optimal shape	0.25	0.08
Reduction	25%	36%
Dynamic case		
Uniform beam	0.664	0.253
Optimal shape	0.421	0.1797
Reduction	36%	29%

The computer programs used in this section are MOLC and MOLD for the concentrated load and distributed load cases, respectively, and can be found on the CD. In the method of lines as applied here, the spatial derivatives are removed using finite difference methods. The resulting *system* of ordinary differential equations is passed to the Matlab ODE solver ODE45. The code of Professor Schiesser contains some excellent graphics. The reader has only to remove the *comments* to access it.

For completeness, a typical set of MATLAB code for the method of lines is listed below. It consists of a *driver* and a subroutine, Prog24.m and Prog25.m. The following routine is the driver:

```
% Program pde_12_main.m (Concentrated Load Case)
% Clear previous files
  clear all
  clc
%
% Parameters shared with the ODE routine
  global ncall n alp bet
  n=21;
  alp=-.5;
  dalp=.5;
  dbet=.2;
  nstep1=0;
  for nstep=1:4
      bet=-.2;
      alp=alp+dalp;
  for mstep=1:4
      bet=bet+dbet;
      nstep1=nstep1+1
%
% Initial condition
  for i=1:(n-1)*2
    u0(i)=0.;
  end
%
% Independent variable for ODE integration
  t0=0.0;
  tf=8.;
  np=100;
  tout=linspace(t0,tf,np);
  nout=n-1;
  ncall=0;
%
% ODE itegration
      [t,y]=ode45(@pde_1,tout,u0);
% Store numerical solution
  %Plot displacement at top
  for i=1:np
    u_plot(i)=y(i,(n-1)*2);
  end
  num1=max(u_plot);
  num2=min(u_plot);
  A(nstep1,1)=alp;
  A(nstep1,2)=num1;
  A(nstep1,3)=num2;
  A(nstep1,4)=bet;
  fprintf('\n ncall = %4d\n',ncall);
%
% Plot numerical solution at top
 % figure(1)
  %plot(t,u_plot); axis tight
%  $title('u(top) vs t'); xlabel('t'); ylabel('u(top,t)')
%
% Plot numerical solution in 3D perspective
```

```
%figure(2);
%colormap('Gray');
%C=ones(n-1,np);
%g=linspace(0,1,n-1); % For distance x
%for i=1:np
%    for j=1:n-1
%        v(i,j)=y(i,j+n-1);
%    end
%end
%h1 = waterfall(t,g,v',C);
%axis('tight');
%grid off
%xlabel('t, time')
%ylabel('x, distance')
%zlabel('u(x,t)')
%s1 = sprintf('Beam Equation - MOL Solution');
%sTmp = sprintf('u(x,0) = 0');
%s2 = sprintf('Initial condition: %s', sTmp);
%title([{s1}, {s2}], 'fontsize', 12);
%v = [0.8616   -0.5076    0.0000   -0.1770
%      0.3712    0.6301    0.6820   -0.8417
%      0.3462    0.5876   -0.7313    8.5590
%           0         0         0    1.0000];
%view(v);
%rotate3d on;
end
end
save A;
```

The following *subroutine* defines the functions α and β and introduces the finite difference approximations of the spatial derivatives:

```
function yt=pde_1(t,y)
% Distributed Load modified for conc load
% Problem parameters...nx-1 points (actually, 2(nx-1))
  global ncall n alp bet
  nx=n;
  xl=0.0;
  xu=1.0;
%
% PDE
  dxx=(xu-xl)/(nx-1);
  dx2=dxx^2;
  dx3=dxx*dx2;
  dx4=dx2^2;
  for i=1:nx-1
  x=(i)/(nx-1);
  ak=1.-alp*(x-.5)+bet*(-1/3+(1-x)^2);
  akp=-alp-2.*bet*(1.-x);
  akpp=2.*bet;
%
%   uxxxx

    if(i==2)                    uxxxx(i)=(y(i+2+nx-1)-4.0*y(i+nx+1-
1)+6.*y(i+nx-1)-4.*y(i+nx-1-1))/dx4;
```

```
        elseif(i==1)               uxxxx(i)=(y(i+2+nx-1)-4.0*y(i+nx+1-
1)+7.*y(i+nx-1))/dx4;
        elseif(i==nx-1)            uxxxx(i)=(2.*y(i+nx-1)-4.*y(i+nx-1-
1)+2.*y(i+nx-2-1)-2.*dx3/ak)/dx4;
        elseif(i==nx-2)       uxxxx(i)=(-2.0*y(i+nx+1-1)+5.*y(i+nx-1)-
4.*y(i+nx-1-1)+1.*y(i+nx-2-1))/dx4;
        else                       uxxxx(i)=(y(i+2+nx-1)-4.0*y(i+nx+1-
1)+6.*y(i+nx-1)-4.*y(i+nx-1-1)+y(i+nx-2-1))/dx4;
        end
  %     uxxx
        if(i==1)          uxxx(i)=(y(i+2+nx-1)-2.0*y(i+nx+1-1)-y(i+nx-
1))/(2.*dx3);
        elseif(i==2)                uxxx(i)=(y(i+2+nx-1)-2.0*y(i+nx+1-
1)+2.*y(i+nx-1-1))/(2.*dx3);
        elseif(i==nx-1)    uxxx(i)=-1./ak;
        elseif(i==nx-2)       uxxx(i)=(-1.0*y(i+nx-1)+2.*y(i-1+nx-1)-
1.*y(i+nx-1-2))/(2.*dx3);
        else                        uxxx(i)=(y(i+2+nx-1)-2.0*y(i+nx+1-
1)+2.*y(i+nx-1-1)-y(i+nx-2-1))/(2.*dx3);
        end
  %     uxx
        if(i==1)        uxx(i)=(y(i+1+nx-1)-2.0*y(i+nx-1))/(dx2);
        elseif(i==nx-1)    uxx(i)=0.;
        else               uxx(i)=(y(i+nx+1-1)-2.*y(i+nx-1)+y(i+nx-1-
1))/(dx2);
        end
            yt(i)=-ak*uxxxx(i)-akp*2.*uxxx(i)-akpp*uxx(i);
            yt(i+nx-1)=y(i);
      end
  %
      yt(n-1)=yt(n-1)*2.;
      yt=yt';
  % increment calls to pd
  ncall=ncall+1;
```

7.2.3 The Work of Connor

There is a long tradition of work in the area of structural motion control that is well represented by the very comprehensive book of Connor (2003). Relative to the work of this chapter, Connor considers the optimal shape of a cantilever beam (building). He starts with the dynamic equations of beam vibrations

$$w-(\varphi y'')''=\ddot{y}$$

For free (w=0), steady-state vibrations ($y=\bar{y}e^{i\omega t}$) the equation to be solved is then

$$-(\varphi\bar{y}'')''=-\omega^2\bar{y}$$

If the assumption is made, on the basis of *optimality criteria arguments*, that in an optimal beam the curvature $\bar{y}''$ is constant, say 1 in this case, the above equation can be integrated as

$$\varphi'' = \omega^2 \bar{y} \quad and \quad \bar{y} = x^2/2 \quad using \quad \bar{y} = \bar{y}' = 0 \quad at \quad x = 0$$

Then

$$\varphi'' = \omega^2 x^2 / 2 \quad \Rightarrow \quad \varphi = \frac{\omega^2}{2}\left(\frac{x^4}{12} + c_1 x + c_2\right)$$

In this type of discussion, the two constants c_1 and c_2 are handled using the base shear of the building and the fact that it is the *shape* of φ that is of interest.

7.2.3.1 Tuned Mass Dampers

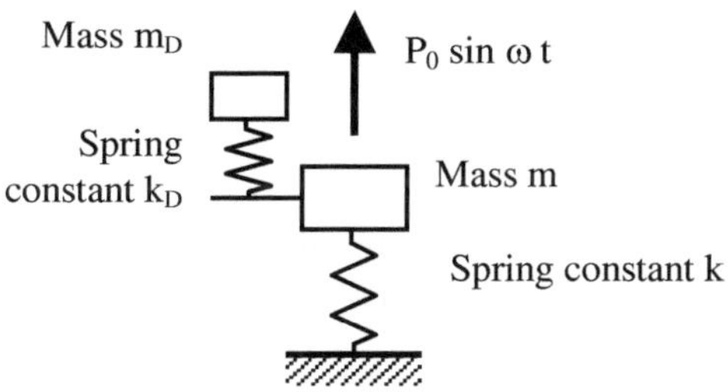

Fig. 7.7. Tuned Mass Damper

Figure 7.7 shows a *tuned mass damper* which is simply a single degree of freedom system (Fig. 7.1) to which a small mass has been added serially. The idea of tuning comes from the fact that if the parameters of the added mass system are selected carefully, a small mass can reduce the amplitude of vibration of the primary mass significantly. Early discussions of the tuned mass damper go back to the work of Den Hartog in the 1930s (Den Hartog 1985); Connor (2003) describes many significant, contemporary applications of its use; and Lee et al. (2006) discuss the optimal design of multiple degree of freedom systems with tuned mass dampers as stochastic systems under wind loads.

Roughly, the equations of motion of the two masses in Fig. 7.7 are

$$P_0 \sin \omega t - ky + k_D(y - y_D) = m\ddot{y}$$
$$-k_D(y - y_D) = m_D \ddot{y}_D$$

For steady-state vibrations, let

$y = a_1 \sin \omega t$
$y_D = a_2 \sin \omega t$

Then a_1 and a_2 must satisfy

$-P_0\, \omega^2 - k\, a_1 + k_D\, (a_1 - a_2) = -m\, \omega^2\, a_1$
$- k_D\, (a_1 - a_2) = - m_D\, \omega^2\, a_2$

The problem is now to select the parameters of the damper, m_D and k_D, to reduce the amplitude of the primary system a_1. With some algebraic effort it can be shown that the added mass is most effective when it is *tuned* so that

$k_D / m_D = k / m$

The reader is referred to the references cited above for discussions of practical problems which, of course, turn out to be more complex.

7.3 Steady-State Problems

This section discusses some problems of the steady-state vibrations of discrete structures (Young and Christiansen, 1966). As a starting point, consider the equations of motion of a discrete system (see also Section 1.11 for the static case)

$$P - K_E \delta = M\ddot{\delta}$$

Here

P – node force matrix
K_E – elastic stiffness matrix
M – mass matrix
δ – node displacement matrix

To get to the equations of free, steady-state vibrations it is only necessary to set P = 0 and $\delta = \bar{\delta} e^{i\omega t}$. What emerges is

$K_E \delta = \omega^2 M \delta$

when the overbar is omitted. Since this equation represents a generalized eigenvalue problem, it is customary to refer to these variables as

δ – eigenvector (displacement)
ω^2 – eigenvalue (frequency squared)

For a comprehensive discussion of the structural eigenvalue problem, the paper of Bathe and Wilson (1973) is excellent. The purpose of this section is simply to collect some results which will be useful later in this section.

If only the fundamental frequency (the lowest eigenvalue) is of interest, the eigenvalue problem is equivalent to the Rayleigh quotient which states that the lowest frequency squared can be obtained as the minimum

$$\omega^2 = \min_{\delta \neq 0} \frac{\delta^T K_E \delta}{\delta^T M \delta} \qquad or \qquad \omega^2 \le \frac{\delta^T K_E \delta}{\delta^T M \delta}$$

The minimum, of course, is achieved when δ corresponds to the first eigenvector, but the inequality is valid for any $\delta \neq 0$. There are several useful features of the eigenvalue problem:

- Scaling: If the structural stiffness matrix is multiplied by a scalar $\alpha > 0$, so are the frequencies: $K_E \rightarrow \alpha K_E \Rightarrow \omega^2 \rightarrow \alpha\,\omega^2$. (This follows directly from the eigenvalue equation).

- Concavity: Beckenbach and Bellman (1971) give in elegant proof of the old result by Courant that ω^2 is a concave function of the matrix K_E or that

$$\omega^2(\alpha K_E' + (1-\alpha)K_E'') \ge \alpha\omega^2(K_E') + (1-\alpha)\omega^2(K_E'')$$

  in which K_E' and K_E'' are two different structural stiffness matrices. (This result follows directly using the Rayleigh quotient.)

- Inverse iteration: Given ω^2, the first eigenvector can be computed conveniently by iterating $K_E\delta^{(n+1)} = \omega^2 M\delta^{(n)}$ (see Bathe and Wilson).

- Homogeneous functions: Let α be a scalar. Any scalar function f of a vector x which has the property $f(\alpha x) = \alpha^n f(x)$ is said to be homogeneous of degree n. It follows that

$$nf = x \cdot \nabla f \qquad and \qquad (n-1)\nabla f = Hx$$

  in which H = Hessian matrix of the function f. Since the structural stiffness matrix K_E: is linear in the member stiffnesses $(K)_I$, scaling implies that ω^2 is homogeneous of degree 1 in K or that $\omega^2 = \nabla\omega^2 \cdot K$. Furthermore, the vector $\nabla\omega^2$ must then be homogeneous of degree 0.

- Derivatives of ω^2 (Fox and Kapoor, 1968): The derivative of ω^2 with respect to any member stiffness, $(K)_I$, can be obtained in a relatively simple manner and is an identifiable structural term. Let the derivative $\partial/\partial(K)_I$ be indicated by a comma. Then

$$(K_E - \omega^2 M)\delta = 0 \quad \Rightarrow \quad (K_{E,I} - \omega^2_{,I} M)\delta + (K_E - \omega^2 M)\delta_{,I} = 0$$

Multiplying by δ now gives

$$\omega^2_{,I} = \frac{\delta^T K_{E,I}\delta}{\delta^T M \delta}$$

This equation can be simplified further. First, it is convenient to normalize with respect to the mass matrix and set $\delta^T M \delta = 1$. Second, it is convenient to decompose the matrix K_E as $K_E = \Sigma_I N_I^T (K)_I N_I$. In this sum each term represents the contribution of member I to the matrix K_E. It follows directly that $K_{E,I} = N_I^T N_I$ and that

$$\partial\omega^2 / \partial(K)_I = \omega^2_{,I} = \Delta_I^2$$

in which $\Delta_I = N_I \delta$ is the member displacement (member length change for the truss). This equation is a simple, but important, result. It identifies the derivative of ω^2 with respect to the member stiffness, $(K)_I$, as equal to the square of the member displacement associated with the normalized eigenvector, δ.

- Concavity and homogeneity: Let f(x) be a concave, homogeneous function of degree 1 so that $f = \nabla f \cdot x$. Furthermore, let x_1 and x_2 be two points on the surface $f(x_1) = f(x_2) = c$ (Fig. 7. 8) and let η be $\perp \nabla f_2$. It follows that since $x_2 = \alpha x_1 + \eta$ and $0 \le \alpha \le 1$, then

$$\nabla f_2 \cdot x_2 = \nabla f_2 \cdot (\alpha x_1 + \eta) \le \nabla f_2 \cdot x_1 \qquad or \qquad \nabla f_1 \cdot x_1 = \nabla f_2 \cdot x_2 \le \nabla f_2 \cdot x_1$$

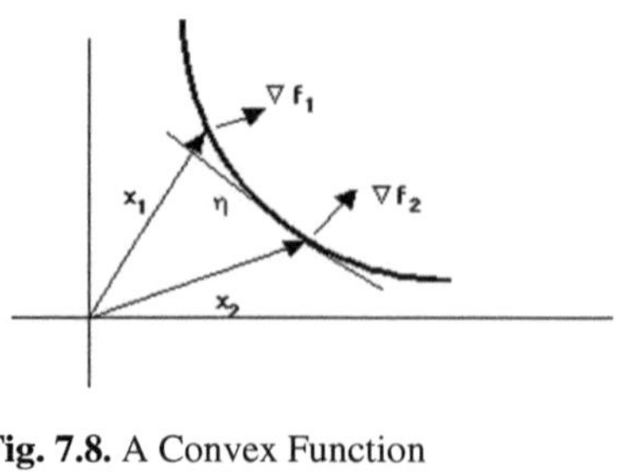

Fig. 7.8. A Convex Function

7.3.1 A Truss Problem

Probably the most simple optimization problem of dynamic structures is the case of a truss with given geometry and mass for which member areas are to be selected so that the weight is minimized while the fundamental (lowest) frequency of vibration, ω, is kept above some specified value. This might be of interest when, for example, rotating machinery is present and an attempt is being made to keep this machinery from exciting the structure supporting it. This problem can be written as find K to

$$\text{minimize} \quad \Delta_a^2 \cdot K \qquad \textit{subject to} \qquad \omega^2(K) \geq c$$

Here $(\Delta_a)_I = \sigma_a L_I / E$ is the "allowable length change" of member I and $K_I = A_I E / L_I$ is the stiffness of member I so that the dot product $K \cdot \Delta_a^2 = \Sigma_I K_I (\Delta_a)_I^2$ is proportional to the material volume. A_I and L_I are the area and length of the Ith truss bar, σ_a is some reference stress, and E is Young's modulus.

Assume that for an optimal solution, the above inequality must be satisfied as an equality. (Otherwise the problem could be scaled for a reduced weight.) The Lagrangian can be formed and an optimality criterion derived as

$$L = \Delta_a^2 \cdot K + \lambda(c - \omega^2) \quad \rightarrow \quad \partial L / \partial K_I = 0 \quad \Rightarrow \quad \Delta_a^2 = \lambda \nabla \omega^2$$

Let f refer to the objective function and $f^* = \Delta_a^2 \cdot K^*$ denote an optimal point. The optimal Lagrange multiplier can be computed as

$$K^* \cdot \Delta_a^2 = \lambda^* K^* \cdot (\nabla \omega^2)^* = \lambda^* c \qquad \Rightarrow \qquad \lambda^* = f^* / c$$

This relationship will be used below to obtain an estimate of the Lagrange multiplier λ.

Before discussing an algorithm to solve this truss problem it can be noted that there are two other related problems:

Problem 1. Given a solution to Eq. (7.1), the solution for some other frequency c′ can be obtained by scaling $K \rightarrow K' = Kc'/c$. (This follows directly from the fact that $\nabla \omega^2$ is homogeneous of degree 0.)

Problem 2. Given the solution to Eq. (7.1), the solution of the problem of maximizing the frequency, given the volume, follows directly by scaling K^* to the proper weight surface.

Let K be arbitrary but scaled so that $\omega^2 = c$ and let K^* be optimal. Then it is possible to select an $\alpha \leq 1$ such that

$$\Delta_a^2 \cdot K^* = \alpha \, \Delta_a^2 \cdot K \leq \Delta_a^2 \cdot K$$

since ω^2 is a concave function of K. Now scale K down to αK which has the same weight as K^*. Since ω^2 is homogeneous of degree 1, it follows that

$$\omega^2(K^*) = \omega^2(K) \geq \alpha\omega^2(K) = \omega^2(\alpha K).$$

7.3.2 An Algorithm for the Truss Problem

The following is an "allowable stress"-type algorithm for the solution of the truss problem of Section 7.3.1. It attempts to find a K^* on the surface $\omega^2 = c$ such that $\nabla\omega_*^2 \sim \Delta_a^2$ iteratively starting with an arbitrary $K_{(0)}$:

Step 1. Given $K_{(n)}$, solve the eigenvalue problem for $\omega^2_{(k)}$ and $\omega^2_{(n)}$ and scale to $K'_{(n)}$ on the surface $\omega^2 = c$ as

$$K'_{(n)} = K_{(n)} \, c/\omega^2_{(n)} \, .$$

Step 2. Compute a new $K_{(n+1)}$ so that

$$(K_{(n+1)})_i (\sqrt{\nabla \omega_*^2})_i = (K'_{(n)})_i (\sqrt{\nabla \omega_{(n)}^2})_i \, .$$

Since $\nabla\omega_*^2$ is unknown, it can be approximated as

$$\nabla \omega_*^2 = \Delta_a^2 / \lambda^* = \Delta_a^2 \; c/\varphi_* \cong \Delta_a^2 \; c/\varphi_{(n)} \, .$$

For more details regarding this algorithm, the reader is referred to Spillers et al. (1981)

7.3.2.1 Motivation for the Algorithm

It is of some interest to compare the above algorithm with the static truss design problem:

Truss design:

Optimality criterion

$$\min |F|\cdot\Delta_a \qquad | \qquad N^T F = P \qquad \Rightarrow \qquad \Delta_i = \text{sgn}(F_i)\,(\Delta_A)_i$$

Iterate $A_i^{(n+1)} = |F_i^{(n)}|/\sigma_a$ or $K_i^{(n+1)}(\Delta_a)_i = K_i^{(n)}|\Delta_i^{(n)}|$

Dynamic problem

Optimality criterion

$$\min \Delta_a^2\cdot K \qquad | \qquad \omega^2 = c \qquad \Rightarrow \qquad \Delta_a^2 = \lambda\nabla\omega^2 \qquad \varphi_{**} = \lambda_* c$$

Homogeneity

$$\omega^2(\alpha K) = \alpha\omega^2(K) \qquad \Rightarrow \qquad \omega^2 - \text{homo}\ 1 \qquad \omega^2 = \nabla\omega^2\cdot K$$

Iterate

$$\Delta_a^2 = \lambda_*\nabla\omega_*^2 \qquad but \qquad \lambda_* = \varphi_*/c$$

$$\sqrt{\Delta_a^2} = \sqrt{\lambda_*}\sqrt{\nabla\omega_*^2}$$

$$K\sqrt{\Delta_a^2} = K\sqrt{\lambda_*}\sqrt{\nabla\omega_*^2}$$

$$iterate\ K^{(n+1)}\sqrt{\nabla\omega_*^2} = K^{(n)}\sqrt{\nabla\omega_{(n)}^2} \qquad with \qquad \nabla\omega_*^2 = \Delta_a^2/\lambda_* \approx \Delta_a^2 c/\varphi_{(n)}$$

The program Prog26.for executes this algorithm:

```
C       PROG26.FOR
C       PLANE TRUSS DYNAMICS
        USE IMSL
        DIMENSION NP(100),NM(100),S(100)
       1 ,AMASS(50,50)
        DOUBLE PRECISION R(100)
        COMMON NP,NM,S,R,AMASS
        COMMON /DAT/E,NB,NN,NS,N,NNN,MAXC
        MAXC=100
C
C       INITIALIZE PARAMETERS/ARRAYS
C
        E = 29.0D06
100     READ(50,150)NB,NN,NS
150     FORMAT (3(I4,3X))
```

```
      WRITE(60,1)NB,NN,NS
    1 FORMAT(I5,'  NO. MEMBERS'/I5,'  NO. NODES'/I5,
     1'  NO.SUPPORTS'//)
      READ(50,156)(R(2*K-1),R(2*K),AMASS(2*K,2*K),K=1,NN)
  156 FORMAT(8X,3F11.6)
      WRITE(60,157)(K,R(2*K-1),R(2*K),AMASS(2*K,2*K),K=1,NN)
 157  FORMAT (1H1,14X,11HCOORDINATES,23X,5HLOADS//
     19X,1HX,9X,1HY,9X,'MASS'//(I4,3F10.0))
      READ(50,151)(NP(L),NM(L),S(L),L=1,NB)
  151 FORMAT(2I5,8X,E10.6)
      WRITE(60,160)(L,NP(L),NM(L),S(L),L=1,NB)
  160 FORMAT(1H1,3X,'MEMBER',5X,'+ END',5X,'- END',6X
     1 ,'AREA'//(3I10,E20.8))
      NNN = NN - NS
      N=2*NNN
      CALL OPT()
      STOP
      END
C
C
C
      SUBROUTINE OPT()
      DIMENSION NP(100),NM(100),S(100)
     1 ,AMASS(50,50)
      DOUBLE PRECISION R(100),UVEC(2),C1,C(100,100)
      COMMON NP,NM,S,R,AMASS
      COMMON /DAT/E,NB,NN,NS,N,NNN,MAXC
      DIMENSION CB(100,100),AN(100,100),DELTA(100),DELA(100)
      DIMENSION ALEN(100),ITYPE(100),EVEC(50,50),EVAL(50)
      DIMENSION XUB(100),XLB(100),ALP(100,100)
     1 ,B(100),DSOL(100),XSOL(100),F(100),BL(100),BU(100)
      SIGA=20000.
      DO 6 I=1,N
      DO 6 J=1,N
    6 IF(I.NE.J)AMASS(I,J)=0.
      DO 7 L=1,NN
      AMASS(2*L-1,2*L-1)=AMASS(2*L,2*L)/386.4
    7 AMASS(2*L,2*L)  =AMASS(2*L-1,2*L-1)
      OMEGA2=1757.
      DO 1 I=1,NB
      DO 1 J=1,N
    1 AN(I,J)=0.
      DO 999 L=1,NB
      K = 2*NP(L)
      M = 2*NM(L)
      CALL UNITV(K,M,C1,UVEC,R)
      ALEN(L)=C1
      DELA(L)=SIGA*C1/E
      IF (K.GT.N) GO TO 2
      AN(L,K-1)=UVEC(1)
      AN(L,K  )=UVEC(2)
   2  IF(M.GT.N) GO TO 999
      AN(L,M-1)=-UVEC(1)
      AN(L,M  )=-UVEC(2)
  999 CONTINUE
C
C
C     SCALE FOR DISPLACEMENTS
C
```

```
      DO 9997 ITER=1,40
      DO 30 I = 1,N
      DO 30 J = 1,N
   30 C(I,J) = 0.
      DO 9999 L=1,NB
      K = 2*NP(L)
      M = 2*NM(L)
      CALL UNITV(K,M,C1,UVEC,R)
      CALL INSERT(C,K,M,UVEC,MAXC,N,E,S(L),C1)
 9999 CONTINUE
      DO 8 I=1,N
      DO 8 J=1,N
C      WRITE(60,*)I,J,C(I,J),AMASS(I,J)
    8 CB(I,J)=C(I,J)
      CALL GVCSP(N,AMASS,50,CB,100,EVAL,EVEC,50)
      WRITE(60,*)EVAL(1),(I,EVEC(I,1),I=1,N)
      RATIOO=OMEGA2*EVAL(1)
      WRITE(60,*)'...SCALING',RATIOO
      PHI=0.
      NB1=NB-1
      DO 88 I=1,NB1
      S(I)=RATIOO*S(I)
   88 PHI=PHI+DELA(I)*DELA(I)*S(I)*E/ALEN(I)
      WRITE(60,*) '...PHI',PHI
      ALAM=PHI/(OMEGA2)
C
C     NORMALIZE EIGENVECTOR
      SUM=0.
      DO 555 I=1,N
  555 SUM=SUM+EVEC(I,1)*EVEC(I,1)*AMASS(I,I)
      BETA=1./SQRT(SUM)
      DO 556 I=1,N
  556 EVEC(I,1)=EVEC(I,1)*BETA
      WRITE(60,611)BETA,(I,EVEC(I,1),I=1,N)
  611 FORMAT(E20.8/(I5,E20.8))
C
      DO 131 I=1,NB
      DELTA(I)=0.
      DO 131 J=1,N
 131  DELTA(I)=DELTA(I)+AN(I,J)*EVEC(J,1)
  61  FORMAT(E20.8/ (I5,2E20.8))
      VOLUME=0.
      DO 71 I=1,NB
      IF(I.NE.NB)VOLUME=VOLUME+ALEN(I)*S(I)
   71 IF(I.NE.NB)S(I)=S(I)*ABS(DELTA(I))/
     1SQRT(DELA(I)*DELA(I)*OMEGA2/PHI)
      WRITE(60,*) '*** VOLUME', VOLUME,ITER
      WRITE(60,611) PHI,(I,S(I),I=1,NB)
 9997 CONTINUE
      RETURN
      END

C
      SUBROUTINE UNITV(K,M,C1,UVEC,R)
      DOUBLE PRECISION R(1),C1,UVEC(2)
      C1=0.
      DO 1 I=1,2
        UVEC(I)=R(K+I-2)-R(M+I-2)
    1 C1=C1+UVEC(I)**2
```

```
            C1=DSQRT(C1)
            DO 2 I=1,2
          2 UVEC(I)=UVEC(I)/C1
            RETURN
            END
C
            SUBROUTINE INSERT(C,K,M,UVEC,MAXC,N,E,S,C1)
            DOUBLE PRECISION C(MAXC,MAXC),UVEC(2),C1
            K1=K
            DO 1 I=1,2
              IF(K1.GT.N) GO TO 1
              M1=K
              DO 2 J=1,2
                IF(M1.GT.N) GO TO 2
                FAC=1.
                IF(I.NE.J) FAC=-1.
                DO 3 L=1,2
                  I1=K1-2+L
                DO 3 L1=1,2
                  J1=M1-2+L1
          3     C(I1,J1)=C(I1,J1)+UVEC(L)*UVEC(L1)*S*E*FAC/C1
          2   M1=M
          1 K1=M
            RETURN
            END
```

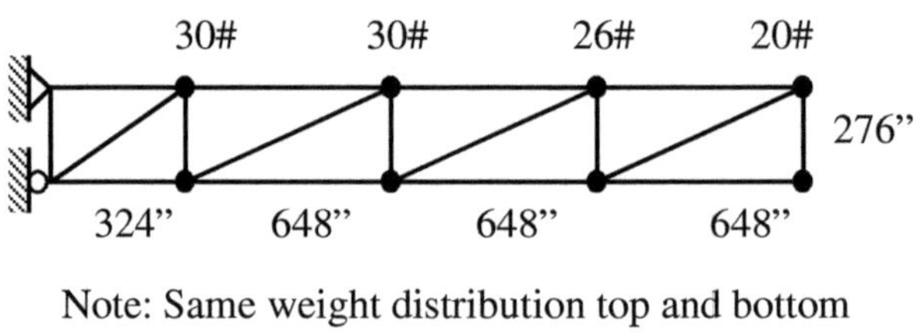

Fig. 7.9. Design Example

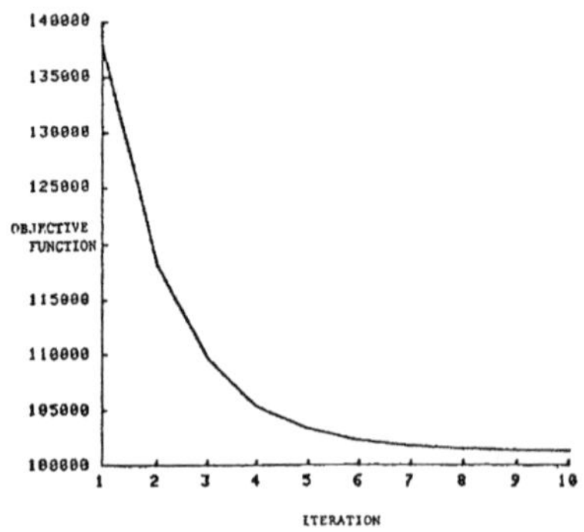

Fig. 7.10. Performance of the Iterative Design Algorithm

Figures 7.9 and 7.10 describe an example using the above algorithm. In this case $c = \omega^2 = 1757\ (\text{rad/s})^2$

7.3.3 An Incremental Solution to the Truss Problem

This is another case in which the incremental approach facilitates a simple solution. The problem stated above is again

$$\text{minimize}\quad \Delta_a^2 \cdot K \qquad \textit{subject to} \qquad \omega^2(K) \geq c$$

where ω^2 satisfies the equation

$$K_E\delta = \omega^2 M\delta$$

When ω^2 and M remain constant, the incremental version of this equation becomes

$$dK_E\delta + K_E\, d\delta = \omega^2 M\, d\delta$$

If this equation is multiplied by δ^T it simplifies considerably as

$$\delta^T (dK_E\delta + K_E\, d\delta) = \delta^T \omega^2 M\, d\delta \quad\Rightarrow\quad \delta^T dK_E\delta = 0 \quad\Rightarrow\quad \Delta^T dK\, \Delta = 0$$

The latter equation can be further decomposed to read

$$\sum_{BARS} \Delta_i^2 dA_i E / L_i = 0$$

The incremental version is then simply

$$\min \sum (\Delta_a)_i^2\, dA_i E / L_i \qquad \textit{subject to} \qquad \sum \Delta_i^2 dA_i E / L_i = 0$$

The program Prog27.for performs this calculation:

```
C       PROG27.FOR
C       PLANE TRUSS DYNAMICS
        USE IMSL
        DIMENSION NP(100),NM(100),S(100)
       1 ,AMASS(50,50)
        DOUBLE PRECISION R(100)
        COMMON NP,NM,S,R,AMASS
        COMMON /DAT/E,NB,NN,NS,N,NNN,MAXC
```

```
      MAXC=100
C
C     INITIALIZE PARAMETERS/ARRAYS
C
      E = 29.0D06
100   READ(50,150)NB,NN,NS
150   FORMAT (3(I4,3X))
      WRITE(60,1)NB,NN,NS
    1 FORMAT(I5,'  NO. MEMBERS'/I5,'  NO. NODES'/I5,
     1'  NO.SUPPORTS'//)
      READ(50,156)(R(2*K-1),R(2*K),AMASS(2*K,2*K),K=1,NN)
 156  FORMAT(8X,3F11.6)
      WRITE(60,157)(K,R(2*K-1),R(2*K),AMASS(2*K,2*K),K=1,NN)
157   FORMAT (1H1,14X,11HCOORDINATES,23X,5HLOADS//
     19X,1HX,9X,1HY,9X,'MASS'//(I4,3F10.0))
      READ(50,151)(NP(L),NM(L),S(L),L=1,NB)
 151  FORMAT(2I5,8X,E10.6)
      WRITE(60,160)(L,NP(L),NM(L),S(L),L=1,NB)
 160  FORMAT(1H1,3X,'MEMBER',5X,'+ END',5X,'- END',6X
     1 ,'AREA'//(3I10,E20.8))
      NNN = NN - NS
      N=2*NNN
      CALL OPT()
      STOP
      END
C
C
C
      SUBROUTINE OPT()
      DIMENSION NP(100),NM(100),S(100)
     1 ,AMASS(50,50)
      DOUBLE PRECISION R(100),UVEC(2),C1,C(100,100)
      COMMON NP,NM,S,R,AMASS
      COMMON /DAT/E,NB,NN,NS,N,NNN,MAXC
      DIMENSION CB(100,100),AN(100,100),DELTA(100)
      DIMENSION ALEN(100),ITYPE(100),EVEC(50,50),EVAL(50)
      DIMENSION XUB(100),XLB(100),ALP(100,100)
     1 ,B(100),DSOL(100),XSOL(100),F(100),BL(100),BU(100)
      DO 6 I=1,N
      DO 6 J=1,N
    6 IF(I.NE.J)AMASS(I,J)=0.
      DO 7 L=1,NN
      AMASS(2*L-1,2*L-1)=AMASS(2*L,2*L)/386.4
    7 AMASS(2*L,2*L)  =AMASS(2*L-1,2*L-1)
      OMEGA2=1757.
      DO 1 I=1,NB
      DO 1 J=1,N
    1 AN(I,J)=0.
      DO 999 L=1,NB
      K = 2*NP(L)
      M = 2*NM(L)
      CALL UNITV(K,M,C1,UVEC,R)
      ALEN(L)=C1
      IF (K.GT.N) GO TO 2
      AN(L,K-1)=UVEC(1)
      AN(L,K  )=UVEC(2)
   2  IF(M.GT.N) GO TO 999
      AN(L,M-1)=-UVEC(1)
      AN(L,M  )=-UVEC(2)
```

```
  999   CONTINUE
C
C
C       SCALE FOR DISPLACEMENTS
C
        DO 9997 ITER=1,40
        DO 30 I = 1,N
        DO 30 J = 1,N
   30   C(I,J) = 0.
        DO 9999 L=1,NB
        K = 2*NP(L)
        M = 2*NM(L)
        CALL UNITV(K,M,C1,UVEC,R)
        CALL INSERT(C,K,M,UVEC,MAXC,N,E,S(L),C1)
 9999   CONTINUE
        DO 8 I=1,N
        DO 8 J=1,N
C        WRITE(60,*)I,J,C(I,J),AMASS(I,J)
    8   CB(I,J)=C(I,J)
        CALL GVCSP(N,AMASS,50,CB,100,EVAL,EVEC,50)
        WRITE(60,*)EVAL(1),(I,EVEC(I,1),I=1,N)
        RATIOO=OMEGA2*EVAL(1)
        DO 88 I=1,NB
   88   S(I)=RATIOO*S(I)
        WRITE(60,*) '...SCALING',RATIOO
C
        DO 131 I=1,NB
        DELTA(I)=0.
        DO 131 J=1,N
  131   DELTA(I)=DELTA(I)+AN(I,J)*EVEC(J,1)
        DO 70 I=1,NB
        ALP(1,I)=(DELTA(I)**2)*E/ALEN(I)
        WRITE(60,*)I,DELTA(I),S(I)
        F(I)=ALEN(I)
        XLB(I)=-.1*S(I)
   70   XUB(I)=.1*S(I)
        BL(1)=0.
        BU(1)=0.
        ITYPE(1)=0
        NROWS=1
C        WRITE(60,*)' *** NROWS',NROWS
        WRITE(60,*)((ALP(I,J),J=1,NB),I=1,NROWS)
        CALL
DLPRS(1,NB,ALP,MAXC,BL,BU,F,ITYPE,XLB,XUB,OBJ,XSOL,DSOL)
        WRITE(60,61)OBJ,(I,XSOL(I),S(I),I=1,NB)
   61   FORMAT(E20.8/ (I5,2E20.8))
        VOLUME=0.
        DO 71 I=1,NB
        IF(I.NE.NB)VOLUME=VOLUME+ALEN(I)*S(I)
   71   IF(I.NE.NB)S(I)=S(I)+XSOL(I)
        WRITE(60,*) '*** VOLUME', VOLUME,NROWS,ITER
 9997   CONTINUE
        RETURN
        END

C
        SUBROUTINE UNITV(K,M,C1,UVEC,R)
        DOUBLE PRECISION R(1),C1,UVEC(2)
        C1=0.
```

```
      DO 1 I=1,2
        UVEC(I)=R(K+I-2)-R(M+I-2)
    1 C1=C1+UVEC(I)**2
      C1=DSQRT(C1)
      DO 2 I=1,2
    2 UVEC(I)=UVEC(I)/C1
      RETURN
      END
C
      SUBROUTINE INSERT(C,K,M,UVEC,MAXC,N,E,S,C1)
      DOUBLE PRECISION C(MAXC,MAXC),UVEC(2),C1
      K1=K
      DO 1 I=1,2
        IF(K1.GT.N) GO TO 1
        M1=K
        DO 2 J=1,2
          IF(M1.GT.N) GO TO 2
          FAC=1.
          IF(I.NE.J) FAC=-1.
          DO 3 L=1,2
            I1=K1-2+L
          DO 3 L1=1,2
            J1=M1-2+L1
    3     C(I1,J1)=C(I1,J1)+UVEC(L)*UVEC(L1)*S*E*FAC/C1
    2   M1=M
    1 K1=M
      RETURN
      END
```

The advantage of the incremental solution is again its generality. That is, most of the time we do not design for simply a single frequency constraint but commonly there are many constraints to be dealt with. With the incremental solution, again, it is a simple matter to add additional constraints as required.

8 Multicriteria Optimization

This chapter discusses an extension of the optimization problem of this text to one in which there is more than one objective function. It includes a discussion of Pareto optimal sets and methods for converting multicriteria problems to conventional optimization problems.

8.1 Introduction

The optimization problem that is used to start this text is find x to

minimize $f(\mathbf{x})$ subject to $\mathbf{g}(\mathbf{x}) \leq 0$

The point here is that the objective function $f(\mathbf{x})$ is a scalar function of an n-vector $\mathbf{x}$ while the constraint vector $\mathbf{g}(\mathbf{x})$ has m components. The fact that $f(\mathbf{x})$ is a scalar allowed, for example, methods of calculus to be used when solving optimization problems.

With multicriteria optimization, the objective function is a *vector*. The multicriteria optimization problem can be written as find **x** to

$$minimize\ f(x) = \lfloor f_1(x), f_2(x), \ldots, f_{n_1}(x) \rfloor \qquad subject\ to \qquad g(x) \leq 0$$

This, first, raises the question of what it means to minimize a vector. Our experience in this case lies with minimizing the *length* of a vector such as the error vector used in the discussion of solving linear equations. While this point will be returned to later, multicriteria optimization more generally has to do with problems where there are different points of view such as architectural design (Gero 1985) where there can be conflicting points of view over questions of structure, building orientation, heating and cooling, site conditions, esthetics, etc. In these cases, it can be difficult to put a number or dollar value on each of these concerns.

There is a famous result from multicriteria optimization due to the economist Pareto. His idea was that you could discuss dissimilar events in the following manner. Say you are at some point x_0 that satisfies whatever constraints need to be satisfied and you would like to find a better point $x_1 = x_0 + dx$. If the point x_1 is better for some of the objective functions and does not hurt the others then the step dx should be made. If it is not possible to find a dx that satisfies these conditions the point x_0 is said to be an element of the *Pareto set*.

W.R. Spillers, K.M. MacBain, *Structural Optimization*,
DOI 10.1007/978-0-387-95865-1_8, © Springer Science+Business Media, LLC 2009

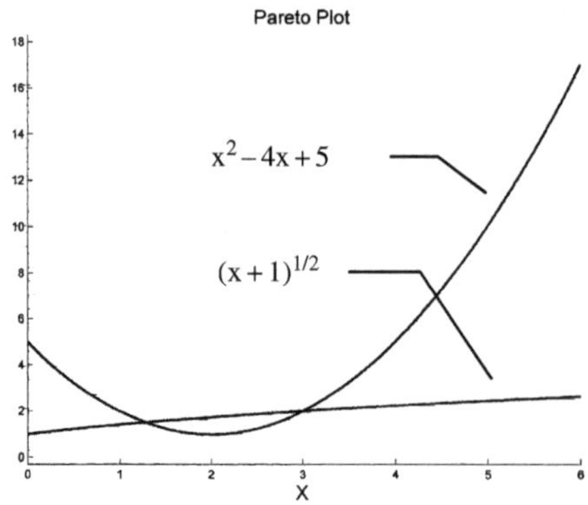

Fig. 8.1. Pareto Plot

Figure 8.1 illustrates the Pareto set. The problem (Ehrgott 2005) is to find x to

minimize $f(x) = [\ (x+1)^{1/2}\ ,\ x^2 - 4\ x + 5\ ]$ subject to $x \geq 0$

If you move in from the right in this figure, both of the objective functions improve until you reach x = 2. To the left of x = 2, the first objective function continues to decrease while the second now begins to increase. The Pareto set is therefore $x \mid 0 \leq x \leq 2$. (Any x in this interval is a member of the Pareto set or called a Pareto optimal solution.)

Sometimes, there is nothing that can be done to resolve the Pareto optimization problem (see Fig. 8.2). For the case of the beam shown, you might ask for dimensions b and d so that you minimize both the center deflection and the volume while keeping the bending stress below some prescribed level. These are

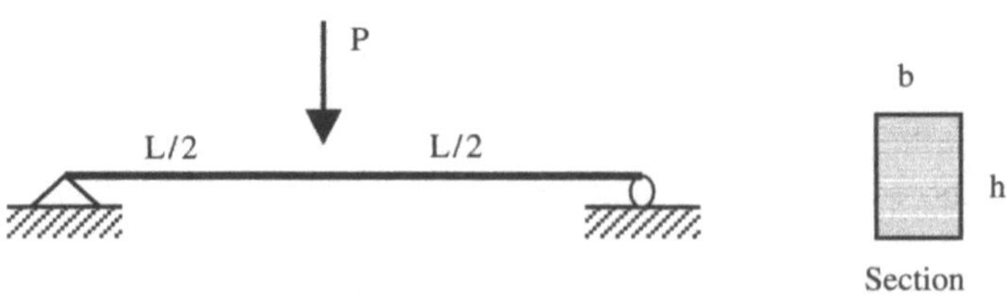

Fig. 8.2. Simply Supported Beam

conflicting requirements since minimizing the volume would tend to produce a small cross section while minimizing the displacement would tend to produce a larger cross section. With regard to detail,

Deflection = $P L^3 / (48 E I)$	$\Rightarrow$	minimize $1/(b h^3)$
Volume = $b h L$	$\Rightarrow$	minimize $b h$
Stress = M / S	$\Rightarrow$	$S \sim b h^2$

The last statement above uses the moment M and the section modulus $S = b h^2 / 6$. If the moment = 50 kip-ft and the allowable stress = 20 ksi, it follows that $b h^2 = 180$.

The analysis of the problem goes something like this. You would like to

minimize [1/(bh3) , bh]   subject to $bh^2 \geq 180$

Let $\eta = bh$. The problem can be written as

minimize [$1/(\eta h^2)$, η]   subject to $\eta h \geq 180$

It can now be seen that (1) $\eta h = 180$ forms a hyperbola in ηh space; (2) to minimize η you can move along this hyperbola to the left making η as small as you like; and (3) along this hyperbola $1/(\eta h^2)$ goes as η which again can be made as small as possible. The result is that there is no Pareto set in this case. Physically, the beam width b gets small while h becomes large minimizing both objective functions.

8.2 Solving Multicriteria Optimization Problems

There are several ways to solve multicriteria optimization problems:

- Construct the Pareto set. This seems to be the case when the issues under discussion are subjective. That is certainly what Gero has done for architectural problems in the book cited above. Ehrgott also offers an extensive discussion of computing the Pareto set. Since the Pareto set can be quite extensive, it is different to deal with than the results of an optimization problem for the structural engineer.
- Reduce the multicriteria optimization problem to a classical single criterion optimization problem. This seems by far to be the most common way of dealing with multicriteria optimization. In fact, one would hazard a guess that this goes on all the time in engineering design. Most engineering design problems have many facets that we manage to include in a single objective function, implicitly or explicitly.

  Formally, it is common to form a kind of *weighted sum* for an objective function

$$f(x) = \sum_{i=1}^{n_1} a_i f_i(x)$$

where the a_i's are weighting coefficients. Doing so, of course, takes you back to classical optimization.

- Using evolutionary algorithms. Fitfler (2001) makes an interesting case for the use of evolutionary algorithms in solving multicriteria optimization problems.

9 Practical Matters: The Work of Farkas and Jarmai

This chapter discusses some practical issues of optimal structural design including sizing the cross section, the effect of materials used, fabrication and painting costs, and a comparison of different sections that are available.

9.1 Introduction

The work of Farkas (1984) and Farkas and Jarmai (1997) is quite exceptional. No one else within the structural optimization literature has managed quite so well to carry optimization technology through to the finished product. (For example, they manage at one point to show the effect of fabrication costs on the design of the member cross section.) They also manage to include a rather nice discussion of structural mechanics within the context of structural optimization. If there is a downside to their work (for us), it is simply that they emphasize European codes rather than American codes.

This chapter looks briefly at some of their results. We at the same time attempt to translate these results into applications familiar to the American reader.

9.2 Sizing Member Cross Sections

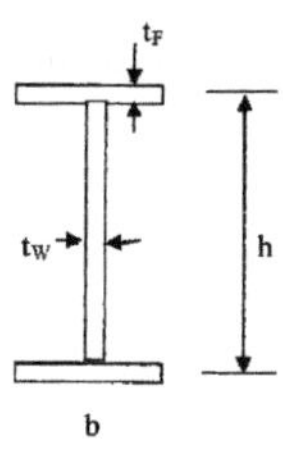

Fig. 9.1. Cross Section of an *I* Beam

Figure 9.1 shows the cross section of a typical *I* beam. In this case, there are four parameters to be determined. The number of parameters can be reduced to two if

W.R. Spillers, K.M. MacBain, *Structural Optimization*,
DOI 10.1007/978-0-387-95865-1_9,

it is assumed that the aspect ratios of the web and flanges are specified typically on the basis of local buckling:

$\beta = t_W / h$
$\delta = t_F / b$

The cross-sectional properties can then be written as follows:

$I = h^3 t_W / 12 + 2 b t_F (h/2)^2 = \beta h^4 / 12 + 2 \delta b^2 (h/2)^2$ (moment of inertia)
$S = I / (h/2) = h^2 t_W / 6 + b t_F h = \beta h^3 / 6 + \delta b^2 h$ (section modulus)
$A = h t_W + 2 b t_F = \beta h^2 + 2 \delta b^2$ (area)

For the case of elastic design it is assumed that there is some given design moment M. If the allowable stress is given, the given moment M implies that the section modulus S is given. The problem then is to find the parameters b and h to minimize the area. That can be done by using the fixed section modulus S to eliminate the term b^2 from the area and then differentiating for the maximum h:

$$A = \beta h^2 + 2(S - \beta h^3/6)/h = \beta h^2 \frac{2}{3} + 2S/h \quad \Rightarrow$$

$$\partial A / \partial h = 0 = \frac{4}{3}\beta h - 2S/h^2 \quad \Rightarrow \quad h = (6S/(4\beta))^{1/3}$$

$$S = \beta h^3/6 + \delta b^2 h \quad \Rightarrow \quad b = (3S/(4\delta h))^{1/2}$$

Farkas and Jarmai make the interesting comment at this point that since plastic design implies a larger aspect ratio for the web thickness, elastic design is advantageous for statically determinate designs.

When deflection considerations control the design, the optimal cross-sectional parameters are different. In this case, the deflection is specified which implies that the moment of inertia I is given rather than the section modulus S. The above analysis can be repeated to give

$$A = \beta h^2 + 4(I - \beta h^4/12)/h^2 = \beta h^2 \frac{2}{3} + 4I/h^2 \quad \Rightarrow$$

$$\partial A / \partial h = 0 = \frac{4}{3}\beta h - 8I/h^3 \quad \Rightarrow \quad h = (6I/(\beta))^{1/4}$$

$$I = \beta h^4/12 + \delta b^2 h^2/2 \quad \Rightarrow \quad b = (I/(\delta h^2))^{1/2}$$

For the case of a box beam, the two webs imply that the above analysis applies when β is replaced by 2 β. In that case

$$\frac{A_{BOX} - A_I}{A_I} = 2^{1/3} - 1 = 0.26$$

or that the box beam has 26% more material than the I beam. Since the area is proportional to $S^{2/3}$ which is proportional to the yield stress $f_y^{-2/3}$ it follows that

$$\frac{A_{36} - A_{51}}{A_{36}} = 1 - (36/51)^{2/3} = 0.2$$

or that increasing the material yield stress from 36 to 51 ksi results in a material saving of 20%.

9.2.1 Effect of Fabrication Costs

The above discussion of an optimal cross section can be extended to include the effect of surface preparation and painting. Farkas and Jarmai use the example in which

$S = 1.404\ mm^3$
$\beta = 1/124$

which gives h = 1377 mm for the case of elastic (stress) design discussed above. They add the cost of painting in the following manner: Let the cost of painting K_p be written as

$K_p = k_p\ (2h + 4\ b)$

where k_p is the cost multiple of the cross-sectional area to be painted. The composite cost of the cross section then becomes

$K = K_m + K_p = k_m\ \rho\ A + K_p$

with

$A = 2\ S / h + 2\ \beta\ h^2 / 3$

(see above). When b = 0.3 h for the painting, it follows that

$$\frac{K}{k_m \rho} = \frac{2S}{h} + 2\beta h^2 / 3 + \frac{3.2 k_p h}{k_m \rho}$$

The optimality condition is now $\partial K / \partial h = 0$. Using the Farkas and Jarmai cost data the optimal h reduces to 1215 mm from 1377 mm described above. That is, considering the painting costs results in a more *squat* cross section.

9.3 Tubular Trusses

Farkas and Jarmai look at a simple truss, Fig. 9.2, as a vehicle to discuss the fact that hollow structural sections (HSS) can be much more efficient than double angles (DA) in some truss designs. The problem in this case is to find the

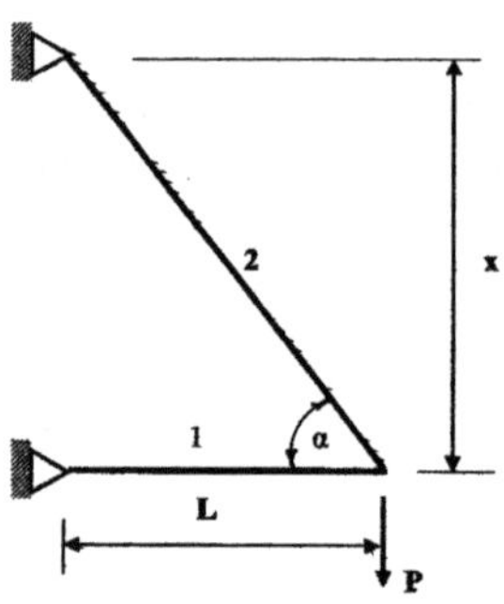

Fig. 9.2. A Simple Truss

optimal angle α when P and L are given:

P = 202 K (factored load)
L = 315 in.
f_y = 36 ksi

Farkas and Jarmai cite some interesting approaches. They use expressions from the European codes for the radius of gyration, r, for double angles as

$$r = a\sqrt{A}$$

with

a = 1.41 for a circular hollow section
a = 0.49 for a double angle cross section

Rather than using the Japanese code as Farkas and Jarmai do, we use the AISC specifications and the EXCEL Solver for a solution. From the AISC specifications

$$F_{cr} = [0.658^{Fy/Fe}]F_y$$
$$F_e = \pi^2 E/(KL/r)^2$$

where

F_{cr} = the compressive strength
F_e = the Euler buckling load
F_y = the yield stress

ht	500		
area1	5		
area2	5		
force1	202		
force2	202		
len2	315		
Fe	36		
Fcr	36		
alp	1		
			=area1*315+area2*len2
		=alp	=ATAN(ht/315)
		=force1	=202/TAN(alp)
		=force2	=202/SIN(alp)
		=Fe	=area1*((3.14159)^2)*29000/((315/(1.41))^2)
		=Fcr	=(0.658^(36/Fe))*36
		=area1	=force1/Fcr
		=area2	=force2/36
		=len2	=315/COS(alp)

Fig. 9.3. Excel Solver Program for the HSS Section

Figure 9.3 shows the EXCEL Solver solution for the hollow circular section. This is Prog33.xls on the CD; the double angle solution is Prog34.xls on the CD. While these results are not identical with those of Farkas and Jarmai they make the same point:

Volume for the double angle solution is 8648 in^3. (α_{max} = 59°)
Volume for the hollow circular section is 5671 in^3. (α_{max} = 55°)
Difference is (8648 – 5671) / 5671 = 0.524 say 52%

The point again is that the HSS sections can be much more effective than double angles in some cases.

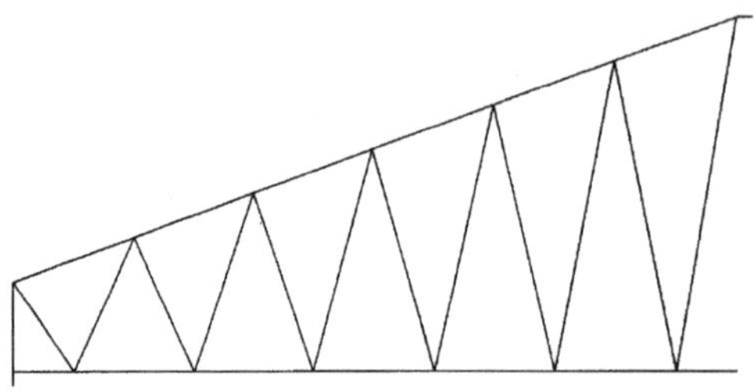

Fig. 9.4. Farkas and Jarmai Truss

9.3.1 The Effect of Shape

As a final example of practical matters, we look at the study of Farkas and Jarmai who do a nice job on the design of a truss with non-parallel chords (Fig. 9.4). This figure shows their optimal design which includes trying different slopes for the upper chord and grouping the members into four groups. This is a very practical approach.

What we have done is to take the optimal shape that he finds and run it through the geometric optimization described in Chapter 3. The optimal design that is produced is shown in Fig. 9.5 and in our terms results in a 9% material saving. In designing this problem we modified the program Prog14.for and now call it Prog28.for which is included on the CD. Doing so required the following steps:

- The objective function: Since this is a statically determinate structure, the constraints remain the same as those used in Chapter 3, but the objective function must be changed to something that reflects the buckling penalty of members in compression. The problem of Chapter 3,

$$\textit{minimize} \sum L_i \,|\, F_i \,| \quad \textit{subject to} \quad N^T F = P$$

$$\textit{becomes}$$

$$\textit{minimize} \sum L_i A_i \quad \textit{subject to} \quad N^T F = P$$

Here the member area A_i (F) is to be computed using the AISC specifications rather than the European codes that Farkas and Jarmai use.

- Compressive strength: The AISC defines the compressive strength F_{CR} as

$$F_{CR} = F_y .658^{F_y / F_e} \quad with \quad F_e = \pi^2 E/(L/r)^2 \ and \ F_e \geq .44 F_y$$

For circular HSS sections, we borrow the result of Farkas and Jarmai that gives the radius of gyration r in terms of the area A as $r = a\sqrt{A}$ with a=1.41. The compressive strength can then be written as

$$F_{CR} = F_y .658^{F_y L^2 / \bar{c} A}$$

with

$$\bar{c} = \pi^2 E a^2$$

- The incremental version: The incremental form of the above optimization problem then becomes

$$minimize \sum A_i dL_i + dA_i L_i \quad subject\ to \quad K_G dR + N^T dF = 0$$

The first term in the objective function has already been discussed in Chapter 3. The second term requires the derivative of A_i with respect to F_i and L_i. This is done by implicit differentiation of the equation

$$F_i / A_i = F_y .658^{F_y L_i^2 / \bar{c} A_i}$$

when $F_i < 0$. This equation plays several roles in the computer program Farkas.for: (1) It is used in the subroutine area to compute the member area A_i given the length L_i and the member force F_i. This subroutine simply performs Newton's method for this purpose. (2) It is differentiated to obtain the derivatives $\partial A_i / \partial F_i \ and \ \partial A_i / \partial L_i$ that appear in the subroutine Xderiv.

With regard to the computer solution shown in Fig. 9.5, we allowed all the nodes to move horizontally as well as vertically. We did this to enable the linear programming algorithm to generate a solution. (Parenthetically, it can happen that the choice of move constraints can preclude a *feasible* solution. When that happens the reader may need to try some trial-and-error approaches when setting up the sequential linear programming algorithm.) Otherwise, there may be some issues concerning the fact that the upper chord is no longer straight, the upper panel points are not equally spaced, and that the members all have different areas. We leave those as open questions noting that automation may mitigate some of them. But it appears that the Farkas and Jarmai design is quite good.

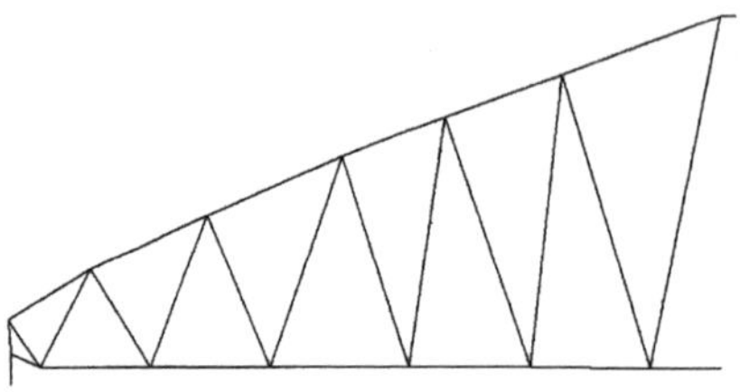

Fig. 9.5. Improved truss

9.4 Problems

1. Compare the designs of Farkas and Jarmai to those obtained using American codes.
2. Discuss the fabrication costs as described in McCormac (2007) and Sarma and Adeli (2000).

10 On Going Work

Structural optimization is a robust area of work and research and for that reason difficult to encompass. This chapter cites some areas of continuing activities within structural optimization that are not included in the main body of this text for several reasons. The easiest of these reasons is complexity: This is the case, for example, of the design of tall buildings where the breadth of activities is remarkable as will be seen below. In other cases like *design theory,* while this activity may be motivated by structural optimization, it is peripheral to it and certainly unpredictable. Finally, the topics included in this chapter reflect the authors' skills or the absence thereof.

In this chapter we offer a light touch giving, at best, a path into certain topics. It would be nice to claim that we are inclusive here but the reader will understand that that is impossible.

10.1 Design of Tall Buildings

We are living in a period of active efforts in tall building design with buildings of record heights constructed on an almost yearly basis. One of the critical phases of the design of a tall building is the design for wind loads. This covers many fields: (1) stochastic optimization because the wind loads can only be described in some probabilistic sense; (2) aerodynamics, certainly, in terms of modeling the wind loads and their interactions with the motion of the structure; (3) dynamic systems, since the structural response is dynamic; (4) structural mechanics in that the interaction of different materials can play a significant role in the structural design of tall buildings; and (5) behavioral psychology since the response of building occupants to structural vibrations is a major concern for the structural engineer. We propose to send the reader to some ongoing studies in these areas.

10.1.1 Wind Loads on Tall Buildings

The design of a monumentally tall building will certainly involve wind tunnel studies and not rely completely on codes for design loads. The paper of Zhou et al. (2003) discusses how wind tunnel studies are carried out and the help that can be obtained from the NatHaz Modeling Laboratory at the University of Notre Dame.

W.R. Spillers, K.M. MacBain, *Structural Optimization*,
DOI 10.1007/978-0-387-95865-1_10, © Springer Science+Business Media, LLC 2009

10.1.2 Tuned Mass Dampers

In Chapter 7 the idea of a tuned mass damper was discussed in connection to its use with a single degree of freedom structure. In fact, today tuned mass dampers are commonly used in tall buildings to mitigate the effects of wind loading (Lee et al. 2006).

It is interesting to note that tuned mass dampers are commonly designed as add-ons. In view of the fact that when you are designing a multi-degree of freedom structure, tuned mass dampers simply become another of many masses to be dealt with; there is an ongoing discussion of whether or not they are really needed (Denoon 2006). Alternatively, a performance-based design would simply specify the desired dynamic characteristics of a design out of which the use or absence of mass dampers would follow directly.

Typically, the discussions of tuned mass dampers start with the equations of motion of a multi-degree of freedom system

$$P - Ky - C\dot{y} = M\ddot{y}$$

where

P – joint node matrix (P contains the disturbing forces)
K – elastic stiffness matrix
C – damping matrix
y – joint displacement matrix
M – mass matrix

The *steady-state* solution involves assuming harmonic forces $P = P_0 e^{i\omega t}$ and a corresponding harmonic response $y = \bar{y}e^{i\omega t}$ that satisfy

$$P_0 - K\bar{y} - iC\bar{y} = -\omega^2 M\bar{y}$$

This equation can then be solved for $\bar{y}$ as a function of the mass matrix M. And if the damping is small, modal analysis can be used effectively. A perturbation analysis can be used to discuss the effect of changing the mass matrix $M \rightarrow M+dM$ as

$$P_0 - K(\bar{y} + d\bar{y}) = -\omega^2 (M + dM)(\bar{y} + d\bar{y})$$

when $C = 0$ or

$$d\bar{y} = (K - \omega^2 M)^{-1} \omega^2 dM\bar{y}$$

This allows the effect of changing the mass on the displacement $d\overline{y}$ to be studied.

10.1.3 Stochastic Processes

With a stochastic process (like wind load on a building) it cannot be said that the load at any point can be described as a function of time. Rather, we are forced to work with some probabilistic parameters of the load such as its average value. While it is, of course, true that none of the properties of any physical system are known precisely, other physical properties such as length or stiffness are typically known to a greater precision than the wind load on a building.

There is a general and ongoing interest in the design of physical systems whose parameters are described probabilistically. Sahinidis (2004) presents a good overview of this work and *opportunities* for additional research.

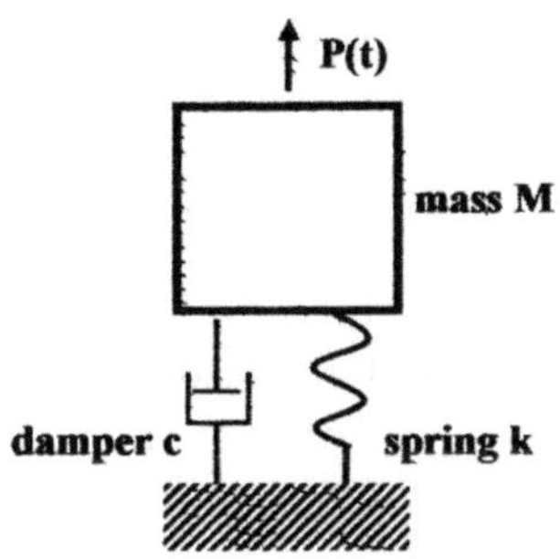

Fig. 10.1. Single Degree of Freedom System

Figure 10.1 will be used to discuss stochastic systems. It is a single degree of freedom system with displacement $y(t)$. The single degree of freedom system has considerable utility given the technique of *modal analysis* that reduces a multi-degree of freedom system to a sequence of single degree of freedom systems.

Very briefly (Newland 1986), some of the notation of stochastic processes will now be listed. The average value, E, of a random variable y is defined to be

$$E[y] = \int_{-\infty}^{\infty} y\, p(y)\, dy$$

where p is the probability density function. It is common to compute the mean-square response

$$E[y^2] = \int_{-\infty}^{\infty} y^2 \; p(y)\, dy$$

in the following manner. The system shown in Fig. 10.1 can be described by the differential equation of motion

$$P(t) - Ky - C\dot{y} = M\ddot{y}$$

For a harmonic input $P = P_0\, e^{i\omega t}$ the frequency response function $H(\omega)$ is defined by

$$y(t) = H(\omega) P_0 e^{i\omega t} \quad \textit{where in this case} \quad H(\omega) = \frac{1}{-M\omega^2 + iC\omega + K}$$

The statistics of the process $P(t)$ are contained within its autocorrelation function R with

$$R_x(\tau) = E[P(t)P(t+\tau)]$$

Finally, the spectral density S is defined as the Fourier transform of the autocorrelation function as

$$S_x(\omega) = \frac{1}{2\pi} \int_{-\infty}^{\infty} R_x(\tau) e^{-i\omega\tau}\, d\tau$$

As a typical result of a stochastic system, Newland computes the mean-square response when the spectral density, S_0, is a constant (white noise),

$$S_x(\omega) = S_0 \qquad \Rightarrow \qquad S_y(\omega) = \frac{S_0}{(k - M\omega^2)^2 + c^2\omega^2}$$

and

$$E[y^2] = \int_{-\infty}^{\infty} \left| \frac{1}{-M\omega^2 + i\omega c + k} \right|^2 S_0\, d\omega = \frac{\pi S_0}{kc}$$

(Notice that the mean-square value of the output, a commonly computed result for stochastic systems, is the integral of the product of the square of the frequency response function and the spectral density.) This should give the reader some idea of the difficulties involved in working with stochastic systems.

In Chapter 7 we looked briefly at the idea of a tuned mass damper that is a 2 degree of freedom system (a single mass system with an attached mass). The question was how to select the parameters of the damper, the stiffness, the attached mass, and the damping to produce an *optimal* system output. In that simple case the steady-state displacement of the primary mass, for example, could be set to zero when the damper was *tuned* to the frequency of the primary system. This is a deterministic and not a stochastic result.

Lee et al. (2006) consider a much more general case. In their case, you start with a system of arbitrary size, add an arbitrary number of dampers, and work with stochastic system input. They start with the general equations of motion for a discrete system

$$P - Ky - C\dot{y} = M\ddot{y}$$

again with

P – joint node matrix (P contains the stochastic disturbing forces)
K – elastic stiffness matrix
C – damping matrix
y – joint displacement matrix
M – mass matrix

Rather than assuming a harmonic input, in the general case it is equivalent to take the Fourier transform of the equation of motion. The equations of motion can be partitioned into three groups of variables:

$$y_r - r \textit{ variables that are to be controlled}$$

$$y_p - p \textit{ variables with stochastic input but not controlled}$$

$$y_e - (n - p - r) \textit{ remaining variables}$$

so that

$$y = \begin{bmatrix} y_r \\ y_p \\ y_e \end{bmatrix}$$

With some algebra, the equations of motion in transform space can be written as

$$\bar{y}_r = \overline{\Phi}(\omega)$$

We will note at this point that it appears that the number of variables (displacements) to be controlled is rarely greater than 2 reflecting the fact that as the number of degrees of freedom increase in a stochastic system, the complexity of the required calculations increases dramatically. It is common in the design of a building for wind loads to assume that the spectral density of the wind is constant and specified by some building code.

In the spirit of the one-dimensional problem above, Lee et al. construct the r × r power spectral density matrix for the output. They then take the integral of the trace of this matrix as a performance index (an objective function) and use Newton's method for an improved solution.

Chan and Chui (2006) carry this design process through to member sizing for a 45-story building. They study the effect of member sizes on the acceleration of the first and second modes of vibration of the structure using an optimality criteria approach to solve the optimization problem. This is an excellent demonstration of an application of the state of the art of structural optimization.

10.2 Heuristic Algorithms

There was a brief mention of genetic algorithms in Chapter 2. More broadly, genetic algorithms fall within the category of evolutionary algorithms where there is considerable development underway. For example, Michalski and Kaufman (2006) discuss what they call a *Learnable Evolution Model* of computational intelligence and encoded expert knowledge. In their system, the random mutations that are typical of genetic algorithms are replaced by "the novel idea of creating new individuals under the guidance of computational intelligence, specifically through an inferential process of hypothesis generation and instantiation". This is an interesting combination of the use of expert systems and genetic algorithms. It responds to two questions regarding the application of genetic algorithms: (1) When you have a large number of parameters to be selected, the random features of genetic algorithms can be time-consuming and hark back to the question of whether enough monkeys with typewriters could eventually reproduce the Bible? and (2) What is an appropriate way to parameterize a complex design situation? One would suspect that eventually, optimization algorithms will combine many or all available approaches.

We commented in Chapter 2 that an advantage of genetic algorithms is the fact that they are not restricted to small changes in a design and can therefore produce *creative* results. An example of this is the work of Hartfield et al. (2007) who have a genetic algorithm sitting on top of existing design modules for the preliminary design of a ramjet-powered missile. Their model has 2 goals, 20 design variables, and

100 population members. It progressed through 40 generations and 40,000 missile designs.

Other current applications of genetic algorithms include Balling et al. (2006) designing the shape and topology of skeletal structures and van de Lindt and Dao (2007) who are designing shear walls for buildings. There is also the notable work of Camp and Bichon (2004) who apply the so-called method of *ant colony optimization* to the design of space trusses.

10.3 Extending the Design Process

Broadly speaking again, while this book looks at rather restricted design situations, it is natural to move on to the automation of *systems* rather than components. That approach is widely used within the aerospace industry. An example of this work comes from Queipo et al. (2005) who develop the idea of a *surrogate* as a means of extending available design information beyond situations for which it has been developed.

Historically, one way to extend the design process involves the computation of system topology, particularly in the case of trusses. Early work in this area took shape optimization routines for trusses and added simple schemes for generating new topologies as described earlier in this text. Even better, the early work of Michell would take a given design situation and produce optimal designs including both member sizes and member topologies. Shape optimization has remained a very robust topic including Allaire et al. (2004) and Du and Qin (2004) who want to *carve* skeletal structures out of solid objects and Hagishita and Ohsaki (2008) who want to *grow* new designs from existing designs.

10.4 Design Theory

The ultimate extension of structural optimization is something called *design theory*. The idea is that as design has gotten more and more formal over the years, people began to ask questions like is there some sort of theory behind the design activities that engineers are so familiar with? Given some sort of design task, we seem to produce artifacts in a rather unstructured manner. Is there some organization to the design process? Are there certain rules that underlie how design should be carried out?

Creativity seems to be a key to these questions. If there are to be creative activities then there cannot be a set of simple rules that designers follow because that would be contrary to what we now regard to be creative behavior. Without answering these questions, we point to some interesting activities that are now ongoing.

10.4.1 Robust Design

The high end of the discussions of engineering design comes out of the Santa Fe Institute. The Santa Fe Institute was created in 1984 around the idea of *complexity*. Complexity was to be the new science. The idea was that the study of complexity would allow us to *understand* everything. While we typically look at and teach design starting with extremely simple objects (as is the case in this text), the designs we create are typically part of an extremely complex system. Complexity theory attempts to study this complex system. What is the theory behind it? Is there a theory behind it? (Jen 2005)

The Internet is commonly offered as the example of engineering design run wild. It started as a simple idea for communications among the military in the United States. It developed into a massive system that, so the story goes, no one understands. It has societal implications that are huge. Its demise could be catastrophic.

Generally, we engineers take a bottom-up approach to our work. That is, we design some object on the basis of simpler designs that have preceded us. As the computer has brought with it the possibility of automation, it has also brought with it the idea of automating the next step in the design process, whatever it is. This led to the idea of design theory. Is there some theory behind the way we work as designers? There is a big gap between the design of some specific object and the worldview into which it eventually fits. The study of complexity would bridge that gap. It might even explain creativity.

10.4.2 Creativity in Design

Once you organize your thoughts on design, the idea of creativity emerges. You can be creative on many levels from solving a problem of crowded reinforcement in a column to designing a plane for man-powered flight. To the extent that creativity represents a laudable human characteristic, it becomes an important area of study. One approach to creativity has been to recognize the fact that we do not understand it and simply try to support it with appropriate tools (Shneiderman 2007).

The support of creativity brings us to the behavioral sciences (Visser 2006). How is the designer to handle the massive amounts of information that are now available to him/her? What is the best way to present information to the designer? What is the best workstation configuration for the designer to use? This is particularly the realm of the psychology of human–computer interaction.

10.4.3 Design Ontologies

Design ontologies represent an attempt to collect the available information we have about design or even structural design. Coming out of database engineering, the construction of design ontologies reflects the fact that as more and more information becomes available to the engineering designer, important questions arise concerning how to organize this information (Brandt et al. 2008).

10.4.4 Architectural Design

Given its breadth, architectural design is particularly interesting in terms of the optimization of the design problem parameters. This breadth can be seen, for example, in the work of Gero and Maher (2005). They have gone from what we would consider to be common architectural details to the most broad discussions of design including imagery, esthetics, visiospatial behavior, psychometrics, etc. The web site listed above certainly deserves a look from anyone interested in the more broad issues of design.

10.5 Available Computational Algorithms

Much of the early work in structural optimization had to do with methods for solving mathematical programming problems. That is, before the engineer could apply optimization technologies, he/she had to worry about how the required computations were to be carried out. Certainly, initially there was not much available for the designer to work with.

Somewhere in the 1990s more algorithms seemed to become available and it is now possible to think in terms of optimization, at least in the long run, as a *utility*, perhaps like the finite element method. In that case, it becomes important for the designer to keep track of developments in mathematical programming. Oberlin and Wright (2006) is a case in point since Wright has been an active worker in this area.

There is also the area of parallel computing (Umesha et al. 2007) that could become important to the designer.

10.4.3 Design Ontologies

A
Using the Computer

This appendix discusses various matters concerned with obtaining computer solutions to problems of structural optimization. The focus of this text has been more on the formulation of a solution rather than the execution and it is thought that the choice of means is largely dependent on what a user is most familiar with. Thus, this section attempts to provide the reader with insight on the practical aspects of computing and the approaches used by the authors. Clearly, tools and methods presented here are not the only way to solve a problem and users are encouraged to determine what works best for their applications.

A.1 Using Computer Languages and Programs

There are several computer languages available. The following is a discussion of what the authors have used extensively and notes to help those using the code on the disk. No comparison is made between languages as this choice is thought to be more of a user preference.

A.1.1 Fortran

Most of the computer programs supplied in this text use the FORTRAN language. All these compile on a PC using Compaq Digital Visual FORTRAN, version 5.0 (1997). This compiler includes the International Mathematical and Statistical Library (IMSL), and some of these IMSL routines are used in our programs. Our limited experience is that they also compile with the new Intel version of FORTRAN. If you are going to purchase a FORTRAN compiler you should make sure that you get the version that includes IMSL routines.

The source code must be compiled to create a functional program. In general the authors compile at the command prompt with the fl32 command. However, it is also possible to set up a workspace using the integrated development environment (IDE) if desired. In fact, the debugger provided in the IDE is particularly useful where one would like to step through the code as it executes. The authors have found that the most direct way to use the FORTRAN compiler is through the command line as

```
fl32  program.for
```

```
program inputfile outputfile
```

When graphics are used, the compile command must be modified to include the appropriate graphics library such as

```
fl32 program.for /libs:qwins
```

For those interested in mixing and matching FORTRAN and C/C++ code including graphics we have used

```
cl /c /MT sdtrpl.cpp
df sdtrpl.obj solplot.for /libs:qwins
sdtrpl
```

A.1.1.1 The IMSL Library

It is our experience that writing a linear programming solver is an order of magnitude more difficult than writing a program to solve linear equations. If that is the case, the engineer needs access to a robust, dependable linear programming solver such as can be found in the IMSL package. (We note that under some circumstances Intel allows free usage of their new FORTRAN compiler http://www.Intel.com/cd/software/products/asmo-na/eng/282048.htm.)

For those who do not have the IMSL routines available, we have put together a library that will provide them when used with a free FORTRAN compiler available on the Internet (also provided on the disk). Users that do not have access to an IMSL-enabled compiler or choose to use another should be able to use any FORTRAN compiler provided that they can correctly access opttools.dll (included on disk). This file is a dynamically linked library that contains the IMSL routines found in this text, and is provided for those without direct access to the IMSL routines used in some of the programs. The procedure may vary for different compilers, so it is outlined here only for the freely available g77 compiler (included in MinGW-5.1.3.exe on the CD):

1) Install g77 by running MinGW-5.1.3.exe (included on CD).
2) Change the name of all IMSL routines in the source code by inserting an "X" at the beginning of the name (e.g., DDLPRS should be changed to XDDLPRS).
3) Compile the main program with the command below. Here a caret (^) is used to indicate the continuation of a single command. This may be typed directly at the DOS-prompt as shown or without the caret on a single line

```
g77 -fno-underscoring -fcase-upper -mrtd ^
    main.for optools.lib -o main.exe
```

This will create the executable program main.exe, which may be executed as usual, by either typing its name at the DOS-prompt or by double-clicking on the program.

A.1.2 Perl

Perl is a very powerful language that is excellent for general-purpose tasks. It is generally not used for computationally intense tasks; however, it has tremendous utility. The authors use it frequently for general tasks that involve text processing and manipulation. For instance, if one were to convert code written in one language to another (e.g., Matlab to FORTRAN), the authors would typically use Perl. Also, it is convenient for more complex batch processing, such as the iterative process of running a program (i.e., any executable), examining the output, and creating a new input file based on the output. Perl is freely available at www.cpan.org.

A.1.3 Makefiles

Large computer programming projects can be difficult to manage. In part, this can be due to the inter-dependence of different modules. When code exists in several different files, one generally wants to only compile those that have changed since the last compilation. Makefiles help accomplish this goal by examining that the time files were last updated and, if needed, performing actions based on user-defined rules. There are many references to this online; a good place to start is http://www.gnu.org/software/make/.

A.2 Matlab

Some of the code included in this text is written for Matlab. While the entire text could have been written using Matlab, the authors found it difficult to discard the historical ties within structural optimization to the FORTRAN language. Further, it is recognized that not all users will have this program. Those who do not have access to Matlab should find it reasonably easy to convert to a compiled language (e.g., FORTRAN). Those that do surely will appreciate the ease that one may generate figures from within the same environment.

A.3 Microsoft Excel

Microsoft Excel is a powerful tool that most engineers are familiar with. Further, it has been used in this text, is commonly found on most desktop computers, and is considered to be the most widely used of tools discussed here. It is assumed that the reader is familiar with the general function of Excel (e.g., formulas and cell references) and two less commonly used features will be discussed here. The advanced user will no doubt already be familiar with both.

A.3.1 The Solver Routine

The "Solver" tool in Microsoft Excel is an excellent device that can be used to solve mathematical programming problems but the novice user must beware of certain details. This appendix describes several steps that the novice should follow. These steps are outlined in the *MBA's Guide to Microsoft Excel.*[1]

Step 1. Verify that the "Solver" is available. This can be done by clicking on the Tools menu item in Excel. If the Solver is not there click on the Add-in menu item to install it (this will generally require the installation disk). In some cases you may need to reinstall Excel to get the Solver.

Step 2. Tell Excel to display formulas rather than results. This can be done using Tools → Options →View → Formulas → OK.

Step 3. Define variable names and give starting values. Put the variables in the upper left corner together with their starting values. Then *select* these items and go to Insert → Name→ Create → OK.

Step 4. Type the components of the mathematical programming problem in cells.

Step 5. Invoke Solver. Tools → Solver

Step 6. Define the problem.

Step 7. Click on "Solve" and then indicate that you want to keep the answer sheet.

---

[1] *MBA's Guide to Microsoft Excel 2002* by Stephen L. Nelson and David B. Maguiness, Redmond Technology Press, Redmond, Washington, 2001.

AN EXAMPLE. This example returns to the two-bar truss problem of Chapter 1. In that case the variables are P = 33 k, B = 30, t = 0.1, E = 3 × 10^4 ksi, and the yield stress = 100 ksi. The problem statement is then

$$\text{Minimize } 0.1885\ d\ (900+h^2)^{1/2}$$

$$\text{Subject to} \quad (900+h^2)^{1/2}/(dh) \leq 0.952$$

$$(900+h^2)^{1/2}/(dh) \leq 352.3\ (d^2 + 0.01)/(900+h^2)$$

The spreadsheet for this problem is shown in Fig. A.1. This just involves entering the terms of the optimization problem in cells in Excel. Fig. A.2 shows the Solver tool of Excel; Fig. A.3 shows the Excel answer sheet for this problem; and Fig. A.4 shows Fox's (1969) representation of the optimization problem. Notice how Fox's solution agrees with the Excel solution. This Excel solution is given on the CD as Program 1.

d	20
h	20
=0.1885*d*SQRT(900+h*h)	
=SQRT(900+h*h)	=0.952*h*d
=SQRT(900+h*h)	=352.3*d*h*(d*d+0.01)/(900+h*h)

Fig. A.1. Excel Spreadsheet

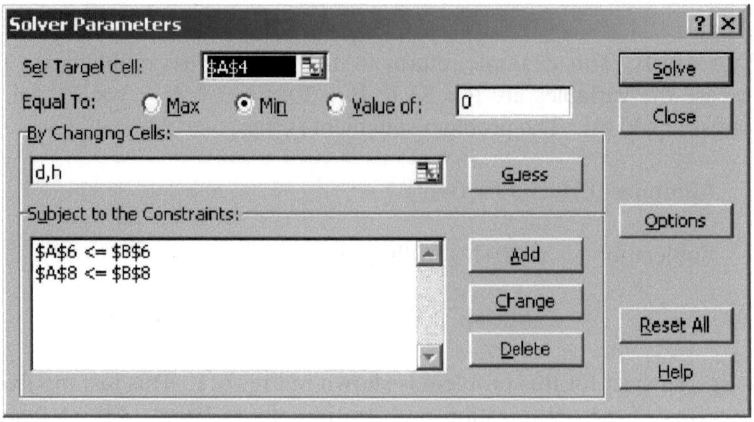

Fig. A.2. Excel's Solver Module

Microsoft Excel 10.0 Answer Report
Worksheet: [Book1]Sheet1
Report Created: 12/13/2007 6:58:39 PM

Target Cell (Min)

Cell	Name	Original Value	Final Value
$A$4	H	135.9292831	12.81319012

Adjustable Cells

Cell	Name	Original Value	Final Value
$B$1	D	20	1.878437574
$B$2	H	20	20.23556957

Constraints

Cell	Name	Cell Value	Formula	Status	Slack
$A$6	H	36.18671408	$A$6<=$B$6	Binding	0
$A$8	H	36.18671408	$A$8<=$B$8	Binding	0

Fig. A.3. Excel's Answer Sheet

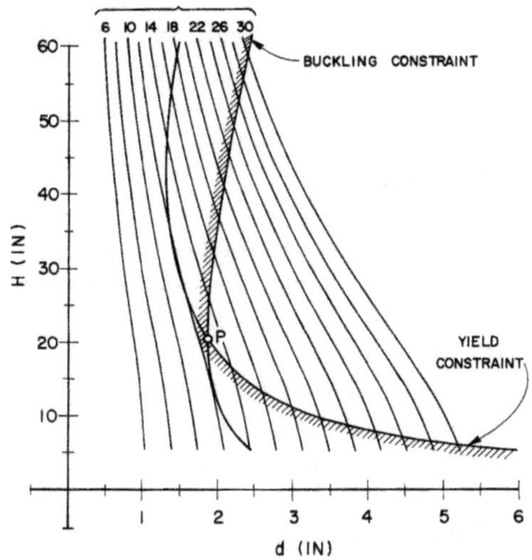

Fig. A.4. Fox's Problem Representation

It may be noted that steps 2 and 3 outlined above are not essential. It is often convenient, however, to make these definitions in order to more easily interpret the solution.

A.3.2 Visual Basic

Excel provides, for automation of tasks via *macros*, a series of commands written in Visual Basic (VB). Typically one would create a macro to automate common tasks (e.g., apply formatting to a cell); however, VB is able to perform any common programming task. There is much help available within the VB editor and recording a macro is often a good way to start. One example of what can be done with macros is included in "tr2d.xls", which will be discussed here.

A.3.2.1 Running the Example

The example spreadsheet will compute the minimum weight of a plane truss subject to allowable stress constraints. Initially the spreadsheet contains no formulas and all computations are done with subroutines. The process is initiated by selecting menu path Tools → Macro → Macros and then running the main subroutine `makeAndOpt`. The process calls the Solver and if this is not properly installed, the routine will not complete (see the notes above on Solver). Select "Yes" when prompted to scale the initial solution and "OK" when prompted to keep the solver solution. The solution should be complete and the minimum weight of 334 should appear on the second sheet, mems.

A.3.2.2 Discussion of the Example

The first two sheets, nodes and mems, are used only to define the initial problem (i.e., input data). The third sheet, system, is a workspace where the system matrix is placed. To understand the process, view the code by opening the VB editor (menu path Tools → Macro → Visual Basic Editor) and navigate to Module 1, where all the code is defined. Some of the functions construct formulas in the worksheets. Formulas are used in this example so that the Solver may understand how the independent variables relate to the objective and constraints (a different approach is discussed later). For example, to programmatically enter a formula in cell B1 on sheet Sheet1 that multiplies the value in A1 by 2, the VB statements are simply

```
Set s = Sheets("Sheet1")
s.Cells(2, 1).Formula = "=A1*2"
```

or even more succinctly in a single statement

```
Sheets("Sheet1").Cells(2, 1).Formula = "=A1*2"
```

Note that VB is an object-oriented language thus the dot-notation is often used. For those unfamiliar with this, the general idea is that objects (e.g., the worksheets above) may contain other objects and/or methods that are referenced with the dot operator. The referenced item may also be an object and thus may itself contain its own objects and methods. All of the functions available in Excel can be referred to in VB. This includes the Solver, which is initialized and executed in the last function, `optimize`.

A final note on the example is about problem size. This example uses the Excel functions MMULT and MINVERSE to solve for the truss displacements. However, this is simply an example and is not the recommended approach for larger systems. It should be clear that to solve a linear system, matrix inversion is not the most efficient method. Fortunately, the Solver routine may be used in a manner that allows for other routines to be called during the solution process. For larger systems, one would include a routine to solve the linear system (e.g., Gaussian elimination) and inform Solver that this routine should be called during the solution. See the parameter StepThru in the Excel help for more on this.

A.4 Freeware

Historically, it has been difficult to obtain a free linear programming package. That may no longer be the case. This section discusses some freeware issues.

In what appears to us to be a new effort, GNU is now offering a free linear programming package at http:www.gnu.org/software/glpk. Given the excellent reputation of GNU, the introduction of this package could be an important event.

The following are some linear programming routines that are available over the Internet:

- LAPKO: This is source code produced by Professor P. Soondal at Seoul National University
  http://www.ICI.ro/camo/lapko.htm
- HOPDM: This is software produced by Professor Gondzio
  http://www.maths.ed.ac.uk/~gondzio/Software/get_hopdm.html
- PCx: This is code made available through the US government
  http://www.mcs.anl.gov/otc/Tools/PCx

A.5 Graphical Interface Applications

Very little has been included in this text in terms of graphical user interface (GUI). It is recognized that for general use, it is often necessary to provide some sort of GUI. The authors have found the Java programming language to be suitable for GUI-based programs. The details are beyond the scope of this text; however, a sample program is included that should give the reader a good starting point. The program will read a text file with geometry and connectivity and render the mesh in a graphic window. The object may be rotated and the node/member numbers may be shown. The sample is found in "View3d.zip" and includes source and a jar-file. The java runtime environment must be installed to run the jar-file. If needed, this can be found at www.sun.com.

B
The Node Method for Trusses

This appendix describes the formulation of the node method of analysis for trusses (Spillers, 1972). It is included here for purposes of reference.

B.1 Introduction

In the context of this book, the truss has a very restricted definition: It is a structure composed of pin-connected elements which is loaded only at its joints. The implication of pin-connected members that are not loaded along their length is well known from elementary mechanics and is simply that the resultant member force lies along a straight line between the ends of the member. Since the line of action of the member force is known, it can be described by

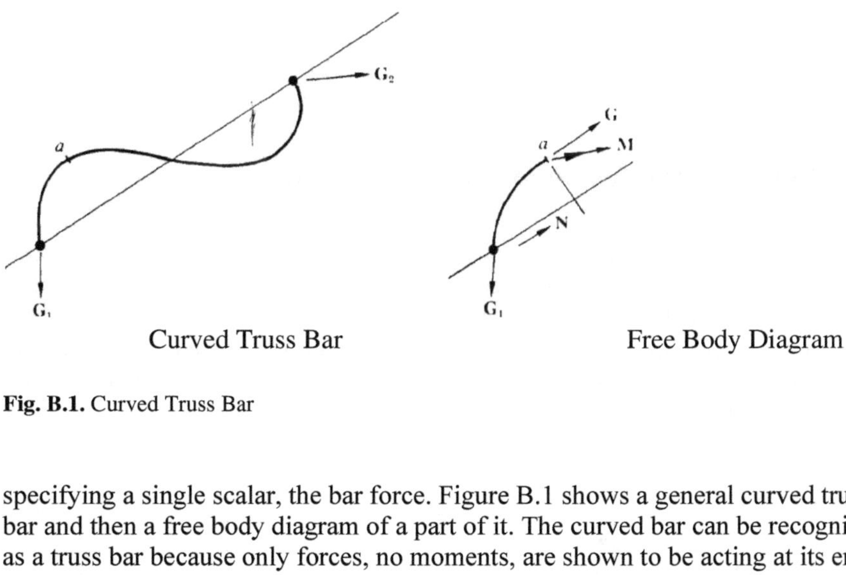

Fig. B.1. Curved Truss Bar

specifying a single scalar, the bar force. Figure B.1 shows a general curved truss bar and then a free body diagram of a part of it. The curved bar can be recognized as a truss bar because only forces, no moments, are shown to be acting at its ends. (By definition, a pin-ended member can have no applied end moments.) Since the sum of the forces acting on the member as a whole must be zero,

$$\mathbf{G}_1 + \mathbf{G}_2 = 0 \qquad \text{or} \qquad \mathbf{G}_1 = -\mathbf{G}_2$$

Let **r** be the position vector of the upper end with respect to the lower end of the member. Taking moments about the lower end gives

$$r \times G_2 = 0$$

from which it follows that since $|r| \neq 0$, when $G_2 \neq 0$, G_2 is parallel to r. The vector G_2 is then a vector whose line of action is known and which can therefore be described as a scalar multiplying either the vector r or a unit vector n.

Figure B.2 shows a structure composed of pin-connected bars subjected to joint loads.

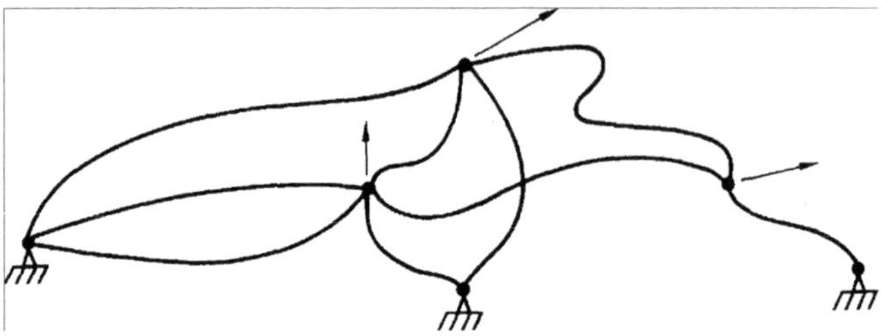

Fig. B.2. A Typical Truss

B.2 A Formal Description of the Truss Problem

The remainder of this appendix is concerned with a rather simple truss problem, the response of a structure composed of straight, pin-connected members subjected only to joint loads. While curved truss members are sometimes useful when, e.g., connections introduce eccentricity, they are not treated here. Such effects as thermal loads, support settlement, and lack of fit are not treated here and are regarded to be "secondary" with regard to the problem solution.

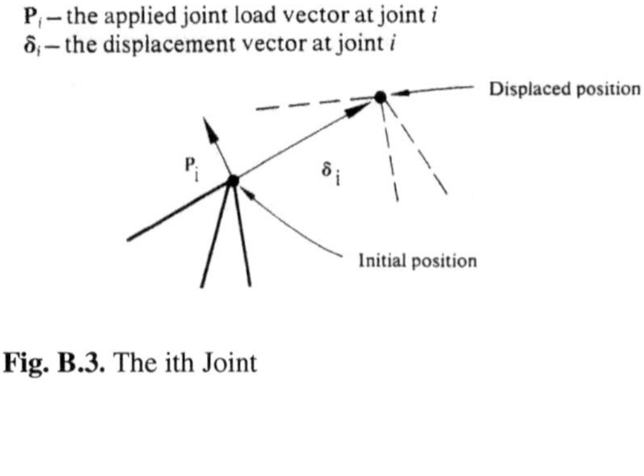

Fig. B.3. The ith Joint

In truss analysis members are idealized as lines which meet at points (nodes) which are called joints. Figure B.3 shows a typical truss joint and the quantities associated with it. For the entire structure these quantities are assembled into the joint load matrix P and the joint displacement matrix δ,

$$\delta = \begin{bmatrix} \delta_1 \\ \delta_2 \\ \cdot \\ \cdot \\ \delta_J \end{bmatrix} \quad \text{and} \quad P = \begin{bmatrix} P_1 \\ P_2 \\ \vdots \\ P_J \end{bmatrix},$$

in which

$$\delta_i = \begin{bmatrix} (\boldsymbol{\delta}_i)_x \\ (\boldsymbol{\delta}_i)_y \\ (\boldsymbol{\delta}_i)_z \end{bmatrix} \quad \text{and} \quad P_i = \begin{bmatrix} (\mathbf{P}_i)_x \\ (\mathbf{P}_i)_y \\ (\mathbf{P}_i)_z \end{bmatrix}.$$

The matrices δ_i and P_i simply contain the components of the vectors δ_i and P_i, respectively; J is the number of movable joints (i.e., joints which are not supports).

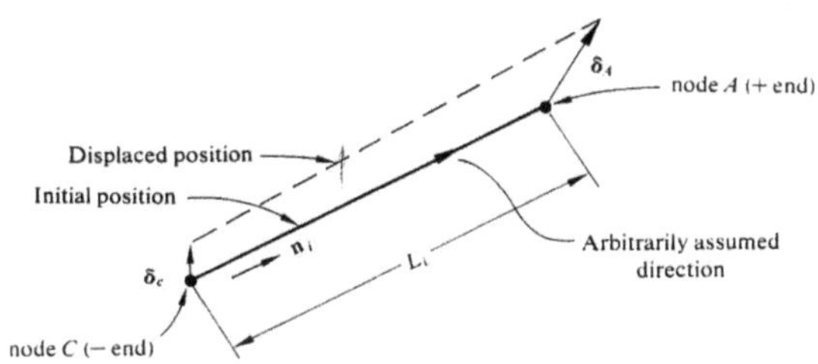

Fig. B.4. Member i

Figure B.4 shows a typical truss bar associated with which is the bar force F_i and the bar length change Δ_i, chosen so that positive F_i and Δ_i correspond to tension or stretching within the bar and a unit vector n_i. Given the displacements of the ends of a bar, it is possible to compute the exact length change of the bar. In linear structural analysis it is customary to use the approximate relationship

$$\Delta_i = n_i \cdot (\delta_A - \delta_C)$$

which uses the notation of Fig. B.4. This involves projecting the joint displacement vectors onto the original position of the bar to obtain the length change. It is only exact when the displaced position of the bar is parallel to its original position but is accurate when they differ by a small rotation. n_i is a unit vector which describes the slope of bar *i* and which has the matrix representation

$$n_i = \begin{bmatrix} (n_i)_x \\ (n_i)_y \\ (n_i)_z \end{bmatrix}$$

For the entire structure the F_i and Δ_i are assembled into the bar force matrix F and the bar displacement matrix Δ,

$$F = \begin{bmatrix} F_1 \\ F_2 \\ \vdots \\ F_B \end{bmatrix} \quad \text{and} \quad \Delta = \begin{bmatrix} \Delta_1 \\ \Delta_2 \\ \vdots \\ \Delta_B \end{bmatrix}$$

where B is the number of bars in the structure.

From elementary mechanics of solids it is known that the bar forces and displacements are related through Hooke's law,

$$F_i = K_{ii}\,\Delta_i \quad \text{in which} \quad K_{ii} = A_i\,E\,/\,L_i$$

where A_i is the area of the ith bar and E is Young's modulus. For the entire structure this may be written as

$$K = \begin{bmatrix} K_1 & & & \\ & K_2 & & \bigcirc \\ & & \ddots & \\ & \bigcirc & & K_B \end{bmatrix}.$$

The matrix K is called the primitive stiffness matrix.

The geometrical bar-displacement–joint-displacement relationship can now be written for the entire structure through the aid of the matrix N, a $B \times J$ (row × column) matrix whose elements N_{ij} are

$$N_{ij} = \begin{cases} \tilde{n}_i & \text{if } j \text{ is the} + \text{end of bar } i \\ -\tilde{n}_i & \text{if } j \text{ is the} - \text{end of bar } i \\ 0 & \text{otherwise} \end{cases}$$

In matrix form the relationship between the member length and the joint displacements can now be written as

$$\Delta = N\delta$$

Note that here it has been convenient to state the definitions in terms of partitioned matrices.

Comparing the definition of N with the definition of the branch node incidence matrix of graph theory provides the motivation for referring to N as a generalized branch node incidence matrix or simply an incidence matrix. Since N contains no columns corresponding to support joints, the implication is that all supports are completely fixed. It may also be noted that any partial support can be replaced by an equivalent fixed support as indicated in Fig. B.5, but care should be taken not to ill condition the system.

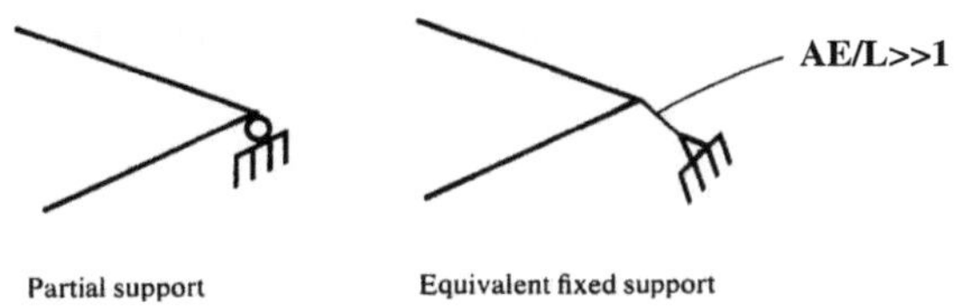

Fig. B.5. Truss Support

To complete the discussion of the node method for trusses, it remains only to discuss the joint equilibrium equations. It will now be shown that they can be written as

$$N^T F = P$$

using matrices previously defined. Figure B.6 shows a typical truss joint whose equilibrium equation (the vector sum of all forces on any joint free-body diagram must be zero) is

$$F_\alpha n_\alpha - F_\beta n_\beta - F_\gamma n_\gamma + F_\delta n_\delta = -P_i \tag{B.1}$$

In general a joint equilibrium equation must contain a term $\pm F_i\, n_i$ for each bar i incident upon the joint — the sign is determined by whether the bar is positively or negatively incident. The node method for trusses then becomes

$N^T F = P$	(node equilibrium equation)
$F = K\Delta$	(Hooke's law)
$\Delta = N\delta$	(branch-displacement–joint-displacement equation)

from which it follows that

$$N^T F = P \quad \rightarrow \quad N^T K\Delta = P \quad \rightarrow \quad N^T KN\delta = P$$

$$or \qquad\qquad (B.2)$$

$$\delta = (N^T KN)^{-1} P$$

This is a system of simultaneous linear algebraic equations on the unknown joint displacements δ; once they have been computed it is a simple matter to compute Δ and F going backward through the system.

The node method begins with forces which satisfy both equilibrium and Hooke's law and writes them in terms of the joint displacements. This insures that the three requirements for solutions are satisfied; in particular, the pieces fit together since their bar length changes are determined from joint displacements.

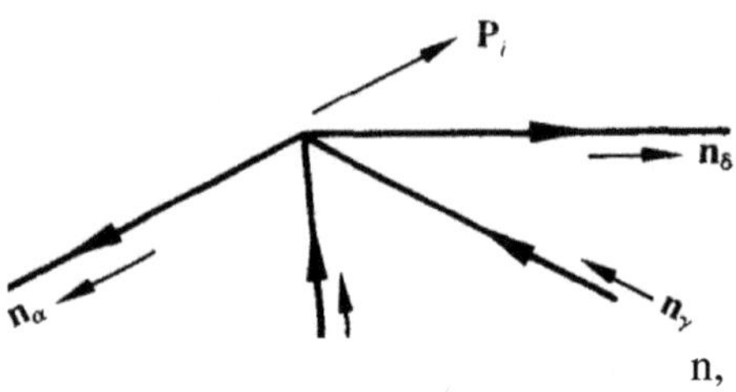

Fig. B.6. Joint Equilibrium

Note that in the case of the equilibrium equations, the omission of support nodes from the matrix N results in not writing equilibrium equations at these nodes.

B.3 A Decomposition

At the heart of the node method is the matrix $N^T KN$, the *system matrix*. While the formal description of the node method for trusses is complete as it is given in Section B.3, from a practical point of view it would be a mistake to program this formulation directly, largely because the matrices N and K are sparse. In this section, a method will be given by which the system matrix can be formed directly from the unit vectors n_i, and the bar stiffnesses K_i without explicitly forming the matrices N and K.

To achieve this end it is convenient to partition the matrix N into its rows

$$N = \begin{bmatrix} N_1 \\ N_2 \\ \vdots \\ N_B \end{bmatrix}$$

Since K is diagonal, it follows that the system matrix can be written as a sum

$$N^T K N = \sum_{i=1}^{B} N_i^T K_i N_i$$

where the term $N_i^T K_i N_i$ can be regarded as the contribution of bar i to the system matrix.

Proceeding in this vein one more step, if neither end of bar i is a support, it follows that N_i contains two non-zero terms and that $N_i^T K_i N_i$ contributes four terms

$$\tilde{N}_i K_i N_i = \begin{bmatrix} \vdots \\ n_i \\ \vdots \\ -n_i \\ \vdots \end{bmatrix} K_i [\cdots \tilde{n}_i \cdots -\tilde{n}_i \cdots]$$

$$= \begin{array}{c} \begin{array}{cc} \text{col. } A & \text{col. } C \end{array} \\ \begin{bmatrix} - & n_i K_i \tilde{n}_i & - & -n_i K_i \tilde{n}_i & - \\ - & -n_i K_i \tilde{n}_i & - & n_i K_i \tilde{n}_i & - \end{bmatrix} \begin{array}{l} \text{row } A \\ \text{row } C \end{array} \end{array}$$

to the system matrix. When either end of bar i is a support node only one diagonal term is generated.

Figure B.7 shows a simple two-dimensional example, which is used here to illustrate the decomposition of the system matrix into the contributions of each of the bars. It is assumed that the stiffness of each of the bars is known; the unit vectors are

$$\mathbf{n}_1 = \frac{1}{\sqrt{2}}\begin{bmatrix}1\\1\end{bmatrix}, \quad \mathbf{n}_2 = \frac{1}{\sqrt{2}}\begin{bmatrix}-1\\1\end{bmatrix}, \quad \mathbf{n}_3 = \begin{bmatrix}1\\0\end{bmatrix}, \quad \mathbf{n}_4 = \begin{bmatrix}0\\1\end{bmatrix}.$$

The matrix N is then

$$N = \begin{bmatrix}N_1\\N_2\\N_3\\N_4\end{bmatrix} = \begin{bmatrix}\tilde{n}_1 & 0\\ \tilde{n}_2 & -\tilde{n}_2\\ 0 & \tilde{n}_3\\ 0 & \tilde{n}_4\end{bmatrix},$$

and the contribution of bar 2, e.g., to the system matrix is simply

$$\tilde{N}_2 K_2 N_2 = \begin{bmatrix}n_2\\-n_2\end{bmatrix} K_2 [\tilde{n}_2 - \tilde{n}_2] = \begin{bmatrix}n_2 K_2 \tilde{n}_2 & -n_2 K_2 \tilde{n}_2\\ -n_2 K_2 \tilde{n}_2 & n_2 K_2 \tilde{n}_2\end{bmatrix}.$$

Since

$$n_2 K_2 \tilde{n}_2 = \frac{K_2}{2}\begin{bmatrix}1 & -1\\-1 & 1\end{bmatrix},$$

it follows that

$$\tilde{N}_2 K_2 N_2 = \frac{K_2}{2}\begin{bmatrix}\begin{bmatrix}1 & -1\\-1 & 1\end{bmatrix} & -\begin{bmatrix}1 & -1\\-1 & 1\end{bmatrix}\\ -\begin{bmatrix}1 & -1\\-1 & 1\end{bmatrix} & \begin{bmatrix}1 & -1\\-1 & 1\end{bmatrix}\end{bmatrix}.$$

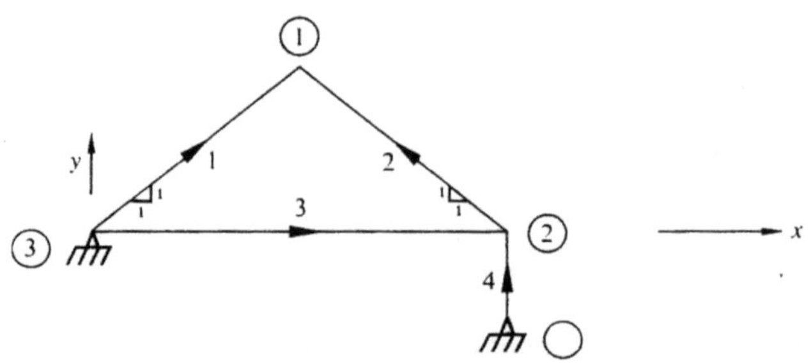

Fig. B.7. A Two-Dimensional Example

Adding the contributions of all four bars, the system matrix for the example is finally

$$\tilde{N}KN = \begin{bmatrix} \frac{K_1}{2}+\frac{K_2}{2} & \frac{K_1}{2}-\frac{K_2}{2} & \frac{K_2}{2} & \frac{K_2}{2} \\ \frac{K_1}{2}-\frac{K_2}{2} & \frac{K_1}{2}+\frac{K_2}{2} & \frac{K_2}{2} & \frac{K_2}{2} \\ \frac{K_2}{2} & \frac{K_2}{2} & \frac{K_2}{2}+K_3 & \frac{K_2}{2} \\ \frac{K_2}{2} & \frac{K_2}{2} & \frac{K_2}{2} & \frac{K_2}{2}+K_4 \end{bmatrix}$$

The following program illustrates the node method for the case of a plane truss. It is listed on the CD as Program 12:

```
C       PROGRAM 3. TR2D.FOR
C       PLANE TRUSS ANALYSIS
        DIMENSION NP(100),NM(100),S(100)
        DOUBLE PRECISION R(100),P(100),C(100,100),UVEC(2)
       1,C1,D1,F1,F2,FAC
        MAXC=100
C
C       INITIALIZE PARAMETERS/ARRAYS
C
        E = 29.0D06
100     READ(50,150)NB,NN,NS
150     FORMAT (3I5)
        WRITE(60,1)NB,NN,NS
      1 FORMAT(I5,'  NO. MEMBERS'/I5,'  NO. NODES'/I5,
       1'  NO.SUPPORTS'//)
        READ(50,156)(R(2*K-1),R(2*K),P(2*K-1),P(2*K),K=1,NN)
    156 FORMAT (8X,4F11.6)
        WRITE(60,157)(K,R(2*K-1),R(2*K),P(2*K-1),P(2*K),K=1,NN)
157     FORMAT (1H1,16X,11HCOORDINATES,28X,5HLOADS/
       1 14X,1HX,17X,1HY,16X,2HPX,16X,2HPY//(I4,4D18.8))
        NNN = NN - NS
        N=2*NNN
C
C       SET UP SYSTEM MATRIX
C
        DO 30 I = 1,N
        DO 30 J = 1,N
     30 C(I,J) = 0.
        WRITE(60,159)
        DO 999 L=1,NB
        READ(50,151)NP(L),NM(L),S(L)
        WRITE(60,160)L,NP(L),NM(L),S(L)
151     FORMAT (2I5,8X,E10.6)
160     FORMAT (3I10,E20.8)
        K = 2*NP(L)
        M = 2*NM(L)
        CALL UNITV(K,M,C1,UVEC,R)
        CALL INSERT(C,K,M,UVEC,MAXC,N,E,S(L),C1)
```

```
  999  CONTINUE
C
C      SOLVE FOR JOINT DISPLACEMENTS
C
       M = N - 1
       DO 17 I = 1,M
       L = I + 1
       DO 17 J = L,N
       IF(C(J,I)) 19,17,19
   19 DO 18 K = L,N
   18 C(J,K) = C(J,K) - C(I,K)*C(J,I)/C(I,I)
       P(J) = P(J) - P(I)*C(J,I)/C(I,I)
   17 CONTINUE
       P(N) = P(N)/C(N,N)
       DO 20 I = 1,M
       K = N - I
       L = K + 1
       DO 21 J = L,N
   21 P(K) = P(K) - P(J)*C(K,J)
       P(K) = P(K)/C(K,K)
   20 CONTINUE
       WRITE(60,161)(I,P(2*I-1),P(2*I),I=1,NN)
  161 FORMAT (1H1,13HDISPLACEMENTS/20X,1HX,19X,1HY//
     1//(I10,2D20.8))
       WRITE(60,162)
  162 FORMAT(1H1,3X,6HMEMBER,9X,2HDL,17X,5HFORCE,
     1 14X,6HSTRESS//)
C
C      COMPUTE MEMBER FORCES AND DISPLACEMENTS
C
       DO 998 I=1,NB
       K = 2*NP(I)
       M = 2*NM(I)
       CALL UNITV(K,M,C1,UVEC,R)
       K1=K
       D1=0.
       FAC=1.
       DO 997 J=1,2
       IF(K1.GT.N) GO TO 996
       D1=D1+FAC*(P(K1-1)*UVEC(1)+P(K1)*UVEC(2))
  996  FAC=-1.
       K1=M
  997  CONTINUE
       F1=D1*E*S(I)/C1
       F2=F1/S(I)
       WRITE(60,1000) I,D1,F1,F2
  998 CONTINUE
       STOP
 1000 FORMAT (I10,3D20.8)
  159 FORMAT (1H1,3X,6HMEMBER,5X,5H+ END,5X,5H- END,6X,4HAREA//)
       END
C
       SUBROUTINE UNITV(K,M,C1,UVEC,R)
       DOUBLE PRECISION R(1),C1,UVEC(2)
       C1=0.
       DO 1 I=1,2
       UVEC(I)=R(K+I-2)-R(M+I-2)
    1 C1=C1+UVEC(I)**2
```

```
      C1=DSQRT(C1)
      DO 2 I=1,2
    2 UVEC(I)=UVEC(I)/C1
      RETURN
      END
C
      SUBROUTINE INSERT(C,K,M,UVEC,MAXC,N,E,S,C1)
      DOUBLE PRECISION C(MAXC,MAXC),UVEC(2),C1
      K1=K
      DO 1 I=1,2
      IF(K1.GT.N) GO TO 1
      M1=K
      DO 2 J=1,2
      IF(M1.GT.N) GO TO 2
      FAC=1.
      IF(I.NE.J) FAC=-1.
      DO 3 L=1,2
      I1=K1-2+L
      DO 3 L1=1,2
      J1=M1-2+L1
    3 C(I1,J1)=C(I1,J1)+UVEC(L)*UVEC(L1)*S*E*FAC/C1
    2 M1=M
    1 K1=M
      RETURN
      END
```

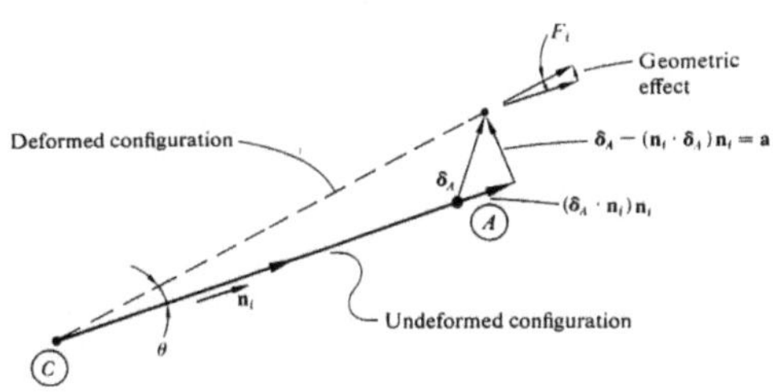

Fig. B.8. Geometric Stiffness Matrix

In linear structural analysis the equilibrium equations are written in the undeformed configuration. The classical treatment of overall truss buckling (not member buckling), on the other hand, adds to the linear formulation. This is a nonlinearity which approximates the effect of small changes of geometry on the linear formulation in a manner similar to the method used to derive the linear response of a membrane or string. How this can be done is now indicated in Fig. B.8 for a single bar, one end of which is allowed to move. The general case is simply a composite in which each end of each bar is considered.

Figure B.8 shows bar i in its undeformed and deformed configurations. Assuming that the bar force F_i remains approximately constant, the displacement δ

generates a force normal to the undeformed position of the bar. Following the string problem, this force has a magnitude of $F_i \times \theta$. In vector form this force is

$$F_i \frac{a}{|a|} \frac{|a|}{L_i} = \frac{F_i}{L_i} [\delta_A - (n_i \cdot \delta_A) n_i]$$

It only remains to introduce this term in an appropriate manner into the stiffness matrix.

C
Convex Sets and Functions: Homogeneous Functions

This appendix reviews some results on convexity and homogeneous functions.

C.1 Convex Sets and Functions

A *set* is said to be convex if the points along a line joining any two points within the set also lie in the set. In order to formalize this statement, it is first of all necessary to formalize the concept of a line between two points in n-dimensional space. Let $x^1, x^2 \in E^n$. A line through these points can be represented by a linear combination of them. That is, any x^3 defined as

$$x^3 = \alpha x^1 + (1-\alpha)x^2 \qquad -\infty < \alpha < \infty$$

when α is a scalar and x^3 lies on the line that passes through points x^1 and x^2. In order to represent a point *between* x^1 and x^2 it is only necessary to restrict α to the range 0,1. Then $C \subset E^n$ is a convex set if $x^1, x^2 \in C$ implies that

$$x^3 = \alpha x^1 + (1-\alpha)x^2 \qquad \in C \qquad when \qquad 0 \le \alpha \le 1$$

A convex *function* has the property of being overestimated by linear interpolation. If *f(x)* is a convex function, then

$$f(\alpha x^1 + (1-\alpha)x^2) \le \alpha f(x^1) + (1-\alpha)f(x^2)$$

Lorch [1951] shows that it is equivalent to define convexity in the following manner. Let *f(x)* = const. define a surface. The surface is convex at some point *x* if the Hessian matrix

$$f_{ij}(x) = \frac{\partial^2 f}{\partial x_i \partial x_j} = [G(x)]_{ij}$$

has a positive-definite quadratic form, that is, if

$$\sum_{i,j} f_{ij}(x) z_i z_j \geq 0 \qquad \textit{for any } z.$$

C.2 Homogeneous Functions

A function is said to be homogeneous of degree n if

$$f(\alpha x) = \alpha^n f(x)$$

By direct differentiation[1] it follows that (Euler's theorem)

$$x \cdot \nabla f = nf \qquad \textit{and} \qquad (n-1)\nabla f = x \cdot G$$

where G is again the Hessian matrix of f. Note that in the case of $n = 1$, the matrix G is singular with an eigenvector x.

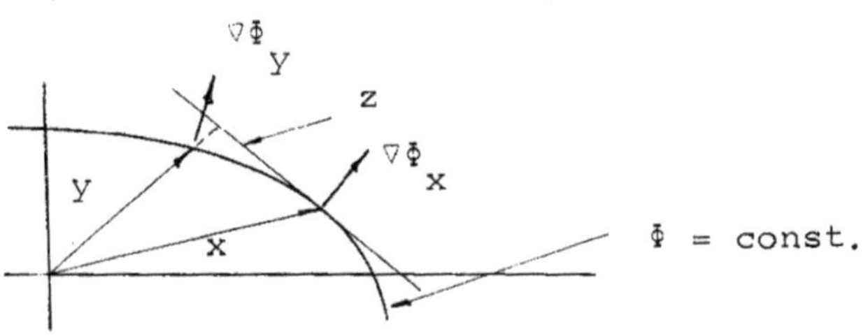

Fig. C.1. Convex Surface

Homogeneity and convexity come together in the following result, Fig. C.1. Let φ be a homogeneous function and let x and y represent two points along a curve φ = const. Then

$$\nabla \phi_y \cdot y = \nabla \phi_x \cdot x$$

---

[1] $$f(\alpha x) = \alpha^n f(x) \quad \Rightarrow \quad f'(\alpha x)x = n\alpha^{n-1} f(x)$$
$$\textit{Let } \alpha = 1 \textit{ then } x\, f' = nf(x) \quad (\textit{scalar case})$$

Let z represent a vector in the plane perpendicular to $\nabla\phi_x$ so that the vector x can be decomposed as

$$x = y/\alpha + z$$

where $\alpha \leq 1$ is a scalar. It follows that

$$\nabla\phi_y \cdot y = \nabla\phi_x \cdot (\frac{y}{\alpha} + z) = \nabla\phi_x \cdot \frac{y}{\alpha} \geq \nabla\phi_x \cdot y$$

or that

$$\nabla\phi_y \cdot y \geq \nabla\phi_x \cdot y$$

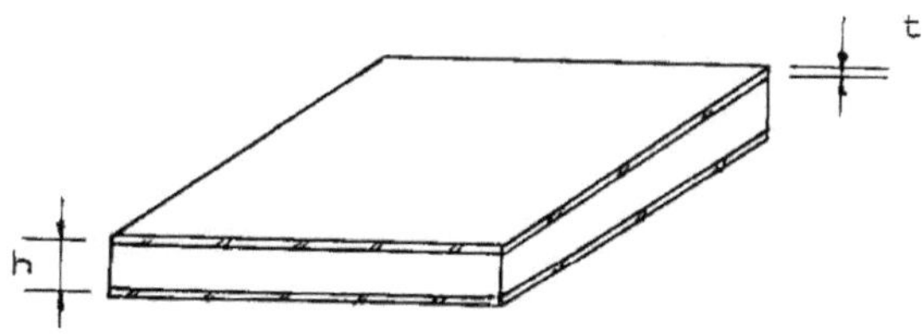

A Sandwich Plate

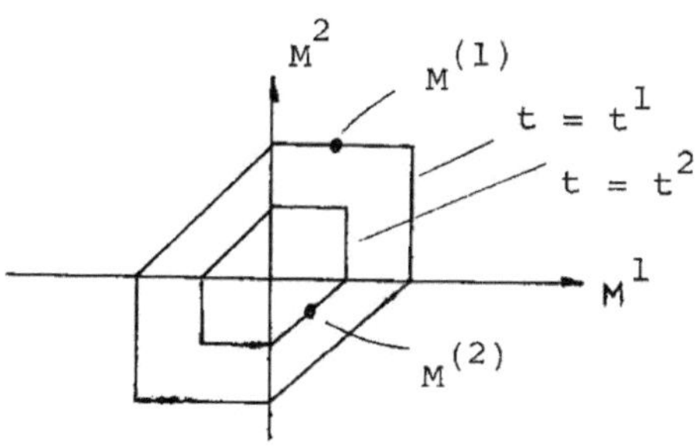

A Convex Function

Fig. C.2. An Example

The sandwich plate, Fig. C.2, illustrates some of these concepts. If its flanges are made of a rigid plastic Tresca material, the required thickness, t, can be expressed in terms of the principal moments M^1, M^2, and the yield stress f_y as

$$t = \frac{1}{f_y h} \begin{cases} \max\left(\left|M^1\right|, \left|M^2\right|\right) & \textit{if } \operatorname{sgn}\left(M^1\right) = \operatorname{sgn}\left(M^2\right) \\ \left|M^1 - M^2\right| & \textit{otherwise} \end{cases}$$

The curves of constant thickness shown in Figure C.2 are all similar since along any ray (a line through the origin) the thickness is linear in the moment. The function t is therefore homogeneous of degree 1 in the moments. It will now be shown that the function t is also convex, i.e.,

$$t[\alpha M^{(1)} + (1-\alpha)M^{(2)}] \leq \alpha t(M^{(1)}) + (1-\alpha)t(M^{(2)})$$

where $M^{(1)}$ and $M^{(2)}$ refer to the points in the figure. Let the curve eb in Fig. C.3 represent the curve $t(\alpha M^{(1)} + (1-\alpha)M^{(2)})$. In view of the linearity of the function t, the chords ad, be, cf are all parallel. Since the triangles

$$\Delta bgc \approx \Delta adc$$

it follows that point g is simply $\alpha M^{(1)} + (1-\alpha)M^{(2)}$. Since the curves of constant thickness are convex, it follows that point g lies inside the curve be which completes the proof.

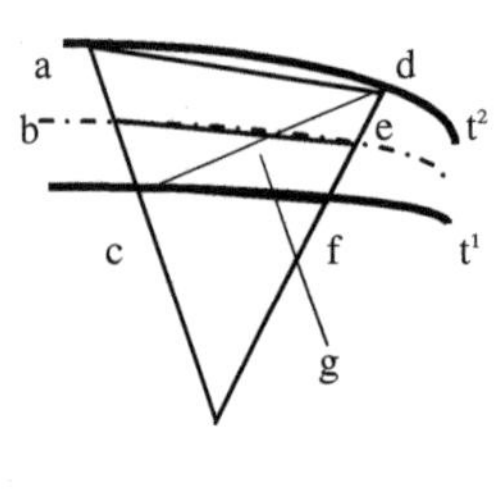

Fig. C.3. Convex Surfaces

D
Structural Optimization Classics

This chapter includes three reprints of papers that are argued here to be *structural optimization classics*. The first is the 1904 paper of Michell that gave rise to the so-called Michell truss; the second is the 1960 paper of Keller on the optimal column; and the third is a paper by Venkayya, Khot, and Berke on the so-called optimality criteria methods.

D.1 Michell Trusses

This is a reprint of the paper

"The Limits of Economy of Material in Frame-structures"
by A. G. M. Michell, *Philosophical Magazine,* S., 6.,
Vol.8, No. 47, Nov. 1904.

It is without peer as it goes from a very general problem statement to a design that includes member layout. The reader will also enjoy Michell's facility with the equations of elasticity.

LVIII. *The Limits of Economy of Material in Frame-structures.*
By A. G. M. MICHELL, *M.C.E., Melbourne* *.

MAXWELL has shown † that for all frames under a given system of applied forces

$$\Sigma \,.\, l_p f_p - \Sigma \,.\, l_q f_q = \mathrm{C}, \quad . \quad . \quad . \quad . \quad . \quad (1)$$

where f_p is the tension in any tie-bar of length l_p, f_q the thrust in any strut of length l_q, and the first sum is taken for all the ties, the second for all the struts. C is shown to be a function of the applied forces and the coordinates of their points of application, and independent of the form of the frame.

Starting from this result, we can find in certain cases lower limits to the quantity of material necessary to sustain given forces, and also assign the forms of frames which attain the limit of economy.

If the greatest tensile stress allowable in the material which is to be employed is P, and the greatest compressive stress Q, the least volume of material in a given frame, consistent with security, is

$$\Sigma \,.\, l_p \frac{f_p}{\mathrm{P}} + \Sigma \,.\, l_q \,.\, \frac{f_q}{\mathrm{Q}} = \mathrm{V}. \quad . \quad . \quad . \quad . \quad (2)$$

* Communicated by the Author.
† Scientific Papers, ii. pp. 175–177.

Now, considering a number of different frames in equilibrium under the same set of external forces, in that one of the frames in which V is least

$$2\mathrm{PQ}\,.\,\mathrm{V}+(\mathrm{P}-\mathrm{Q})\mathrm{C} \quad \text{is also least;}$$

$$i.\,e.,\ 2\mathrm{PQ}\left\{\Sigma\,.\,l_p\frac{f_p}{\mathrm{P}}+\Sigma\,.\,l_q\frac{f_q}{\mathrm{Q}}\right\}+(\mathrm{P}-\mathrm{Q})\{\Sigma\,.\,l_pf_p-\Sigma\,.\,l_qf_q\}$$

$$=(\mathrm{P}+\mathrm{Q})\{\Sigma\,.\,l_pf_p+\Sigma\,.\,l_qf_q\} \quad \text{is least,}$$

or $$\Sigma\,.\,l\,[f] \text{ is least,} \qquad (3)$$

where $[f]$ denotes the numerical value of the force in any bar, and the sum is taken for both struts and ties.

One proof given by Maxwell of the equation (1) above depends on consideration of the virtual work of the applied forces and internal stresses during an imposed uniform dilatation or contraction of the frame. Consideration of a more general type of imposed deformation will furnish information as to the quantity $\Sigma\,.\,l\,[f]$.

Let the space within a given boundary, which encloses a number of different frames subjected in turn to the same set of applied forces, undergo an arbitrary deformation of which the frames partake and such that no linear element in the space suffers an extension or contraction numerically greater than $\epsilon\,.\,\delta l$, where δl is the length of the element and ϵ a given small fraction.

In this deformation, let any bar of length l of one of the frames, A, undergo the small change of length el, which is to be taken as positive when it increases the existing strain in the bar due to the applied forces, negative in the contrary case. The increase in the elastic energy of the bar is elf, and for the whole frame

$$\Sigma\,.\,elf=\delta\mathrm{W},$$

by the principle of Virtual Work, where $\delta\mathrm{W}$ is the virtual work of the applied forces, and is independent of the form of the frame A.

Thus $$\delta\mathrm{W}=\Sigma\,.\,elf \ngtr \Sigma\,.\,[e]l[f]$$
$$\ngtr \Sigma\,.\,\epsilon l\,[f] \ngtr \epsilon\Sigma\,.\,l\,[f]$$

or $$\Sigma\,.\,l_{\mathrm{A}}\,[f]_{\mathrm{A}} \nless \frac{\delta\mathrm{W}}{\epsilon},$$

indicating by the suffixes that the inequality applies to the frame A.

If, however, a frame, M, can be found, such that all its parts have their strains increased equally and by as much as any elementary line in the deformed space, *i. e.* if $e = \epsilon$ in all parts, the signs of inequality may be replaced by that of equality, and

$$\Sigma \,.\, l_M [f]_M = \frac{\delta W}{\epsilon} \not> \Sigma \,.\, l_A f_A,$$

so that $\Sigma \,.\, l_M [f]_M$ is a minimum, and consequently from (3), V_m, the volume of material in the frame M, is also a minimum.

A frame therefore attains the limit of economy of material possible in any frame-structure under the same applied forces, if the space occupied by it can be subjected to an appropriate small deformation, such that the strains in all the bars of the frame are increased by equal fractions of their lengths, not less than the fractional change of length of any element of the space.

If the space subjected to the deformation extends to infinity in all directions, the volume of the frame is a minimum relatively to all others, otherwise it will have been shown to be a minimum only relatively to those within the same assigned finite boundary.

The condition $e = \epsilon$ can evidently be satisfied when all the bars of a frame have stresses of the same sign. The test deformation to be applied is then a uniform dilatation or contraction (according as the frame is in tension or compression) of a space enclosing the frame and extending to any finite boundary or to infinity at pleasure.

The simplest minimum frames of this special class are:—

I. Ties and struts subjected to a single pair of equal and opposed forces.

II. Triangular and tetrahedral frames under forces applied at the angles of the figure, and acting along lines which intersect within the figure.

III. Catenaries in general, the points of application of the applied forces, as well as the forces themselves, being given.

In all these cases the minimum volume of the frame is

$$V_m = \frac{\Sigma \,.\, l[f]}{P} = \frac{\delta W}{\epsilon P} = \frac{\Sigma \,.\, \epsilon F r \cos\theta}{\epsilon P}$$

$$= \frac{\Sigma \,.\, F r \cos\theta}{P},$$

where F is one of the applied forces, r the distance of its point of application, R, from an arbitrary fixed point O,

θ the angle between RO and the direction of F, and P is the allowable stress in the material.

A more general class of frames satisfying the condition $e=\epsilon$, consists of those whose bars, both before and after the appropriate deformation, form curves of orthogonal systems.

In such a system the test strains may be equal and of the same sign, or equal and opposite, in directions at right angles, without the strains in the bars being exceeded by those of any other lines in the field. In the first case the strain is evidently the same in all directions; in the second, if λ, $-\lambda$ are the strains in the principal directions, the strain in any direction at an angle θ to one of them is $\pm\lambda \cos 2\theta = \lambda'$,

$$\therefore [\lambda'] \ngtr [\lambda].$$

The condition that a two-dimensional orthogonal system shall remain orthogonal after equal and opposite extensions of its two series of curves, is that the inclination between any two adjacent curves of the same series is constant throughout their length. This is easily seen as follows:—

Let l be the elementary length, ϕ the change of direction of the curve S_1, between its points of intersection with two

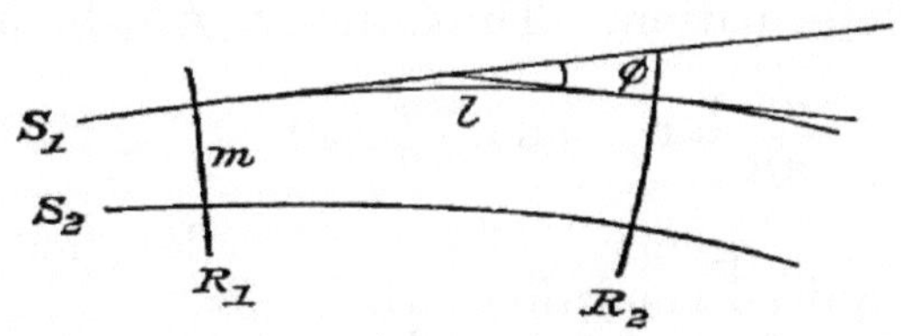

adjacent curves R_1R_2 of the other series, and let m be the elementary length of R_1 between the two curves S_1, S_2 of the S-series, m being taken small in comparison with l.

Let the curves of the S-series be extended by the small fraction λ of their length.

If the element remains rectangular the change in the angle ϕ must be equal to the change in the inclination of the elements R_1 and R_2 to each other,

$$\therefore \delta\phi = \frac{\lambda l - \lambda(l + \frac{dl}{dm} \cdot m)}{m} = -\lambda \frac{dl}{dm} = \lambda\phi.$$

Take now a new element, with sides of comparable length, l, n, formed by the pairs of curves R_1R_2, S_1S_k. Let ψ be the change of direction of R_1 between S_1 and S_k, and let λ' be the extension of the R curves.

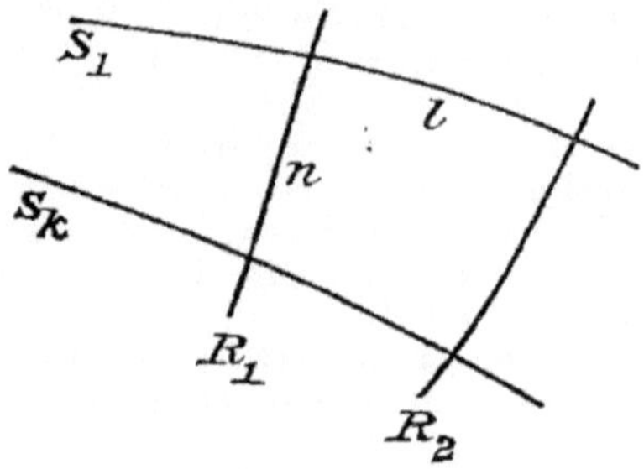

Then, as before, $\delta\psi = \lambda'\psi$; and making a circuit around the element, since the angles remain right angles,

$$\lambda \,.\, \frac{d\phi}{dn} n - \lambda' \,.\, \frac{d\psi}{dl} \,.\, l = 0,$$

but $\frac{d\phi}{dn} \,.\, n - \frac{d\psi}{dl} \,.\, l = 0$, because the angles were right angles before the deformation. Thus since $\lambda' \neq \lambda$, by assumption

$$\frac{d\phi}{dn} = 0, \quad \text{and} \quad \frac{d\psi}{dl} = 0. \qquad \text{Q.E.D}$$

There are two general classes of orthogonal curves satisfying the required conditions, viz. :—

I. Systems of tangents and involutes derived from any evolute curves.

(Since such systems are bounded by the evolute curves, the corresponding frames are in general of minimum volume only relatively to others with the same finite boundaries.)

II. Orthogonal systems of equiangular spirals, with systems of concentric circles and their radii, and rectangular networks of straight lines, as special cases.

Frames whose bars coincide with the curves of any of these systems are therefore frames of minimum quantity of material, for any system of forces consistent with their equilibrium and continuity of displacement.

Frames may also be built up of parts of different systems of these classes provided that continuity of displacement is secured along the lines of junction.

Examples of such frames of minimum quantity of material, for some elementary systems of forces, are given in the accompanying figures.

Compression bars in each case are indicated by thick, and tension bars by fine, continuous lines. Those portions of the lines of principal strain on which material bars are not required are shown by dotted lines.

1. (Fig. 1.) A single force F applied at A, and acting at right angles to the line AB, is balanced by an equal and opposite force and a couple, of moment F × AB, applied at B.

Fig. 1.

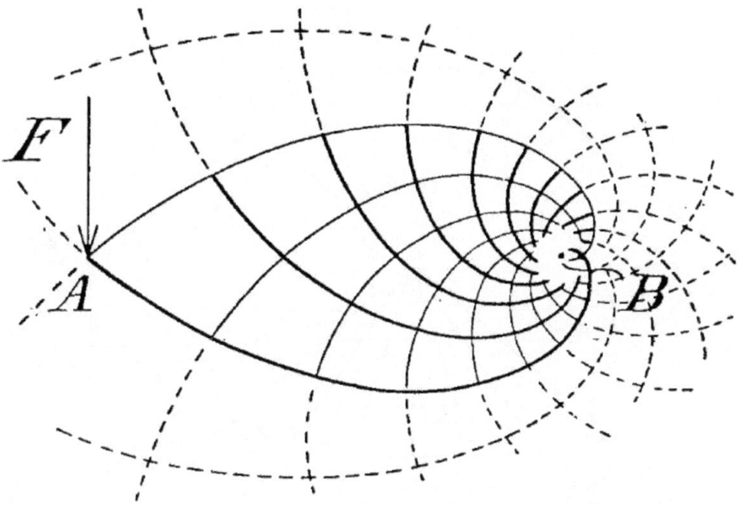

The minimum frame is formed of two similar equiangular spirals having their origin at B and intersecting orthogonally at A, together with all the other spirals orthogonal to these and enclosed between them.

All the spirals of one series are equally extended, those of the other system equally compressed to the same degree. The spirals may be extended to infinity, and the infinite field will remain continuous with equal and opposite strains in the principal directions at all points. The necessary volume of material is infinite if the frame is continued to a mere point at B; but if forces equivalent to the force F, and the couple F × AB at B, are distributed over a small circle with its centre at B, and of radius r_0, the volume necessary is

$$Fa \,.\, \log \frac{a}{r_0} \,.\, \left(\frac{1}{P} + \frac{1}{Q}\right),$$

where $a = AB$, and P, Q are the stresses allowed for tension and compression.

2. *Centrally-loaded Beam.*—(Fig. 2.) A single force, F, is applied at C, the middle point of the line AB, and is balanced by equal parallel forces at A and B.

The minimum frame is composed of the two quadrantal bars DF and EG, having their common centre at C, of all

Fig. 2.

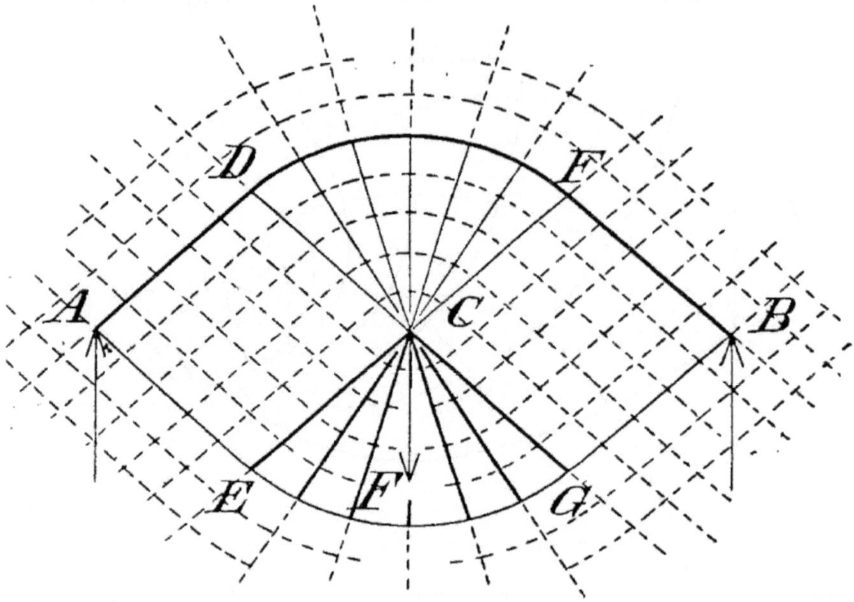

the radii of these quadrants, and their four tangents AD, AE, BF, and BG. The orthogonal system is completed and extended to infinity as indicated by the dotted lines.

The quadrants above C and all straight lines extending obliquely downwards and outwards from the line of action of F are compressed, all the other lines extended.

The necessary volume of material is

$$\mathrm{F}a\left(\frac{1}{2}+\frac{\pi}{4}\right)\left(\frac{1}{\mathrm{P}}+\frac{1}{\mathrm{Q}}\right),$$

where a=AC=CB.

2*a*. (Fig. 3.) With the same applied forces the condition may be imposed that the frame shall lie wholly on one side of the line AB.

The form of the minimum frame is then as in fig. 3, and consists of the semicircle on AB and its radii. The figure to be subjected to the test deformation consists of the semi-infinite plane below AB, all the radii from C being extended and the semicircles about C compressed.

The minimum volume of the frame is

$$Fa \cdot \frac{\pi}{2}\left(\frac{1}{P} + \frac{1}{Q}\right),$$

and therefore exceeds the frame of fig. 2, in which no limitation of extent was imposed, in the ratio $2\pi : (2+\pi)$, or $1{\cdot}22 : 1$.

Fig. 3.

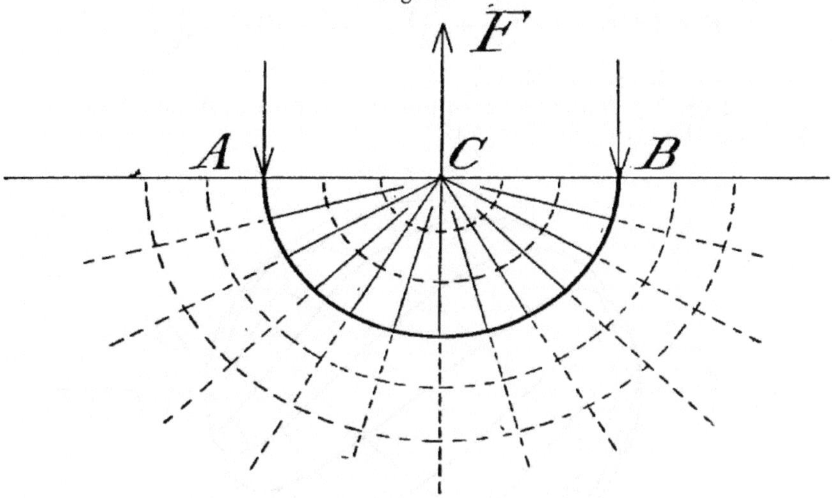

Fig. 4.

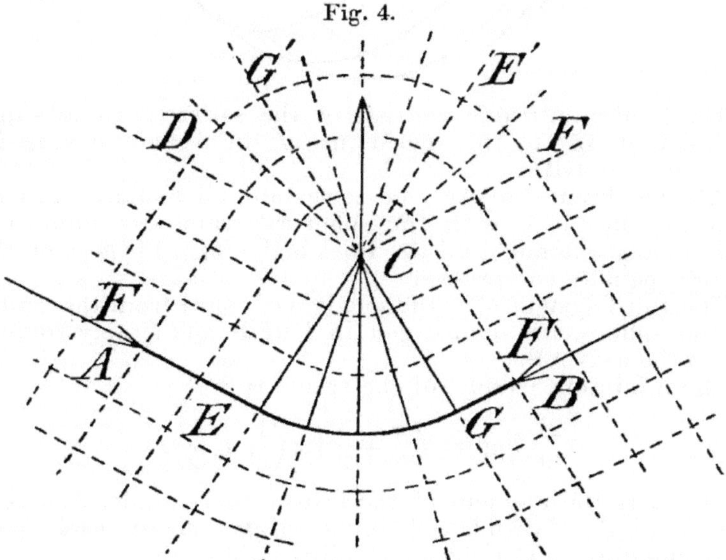

3. (Fig. 4.) Constructions similar to that of figs. 2 & 3 give the minimum frames for three forces, two of which are

equal and directed to a point on the line of action of the third outside the triangle formed by the points of application of the forces. Fig. 4 is an example.

The sectors DCG′, FCE′ undergo uniform bulk compression, the remainder of the field pure shear as in fig. 2.

The volume of the frame is

$$\mathrm{F} \times \left\{ \mathrm{CE} \times \alpha \times \left(\frac{1}{\mathrm{P}} + \frac{1}{\mathrm{Q}}\right) + (\mathrm{AE} + \mathrm{BG}) \times \frac{1}{\mathrm{Q}} \right\},$$

where α is the angle ECG.

4. (Fig. 5.) Equal and opposite couples applied at points A, B on the straight line AB.

Fig. 5.

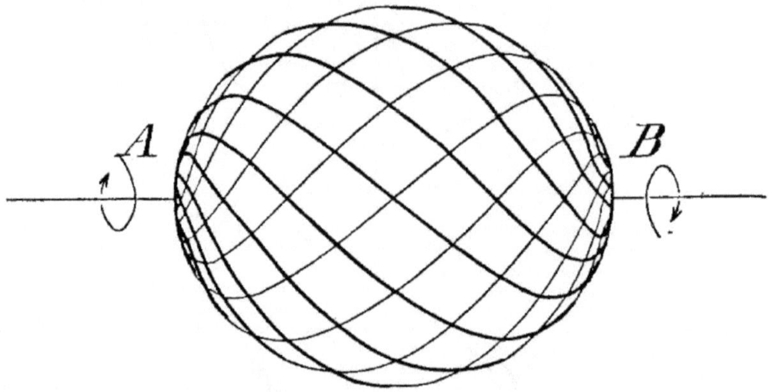

The minimum frame consists of the series of rhumb-lines inclined at 45° to the meridians of the sphere having its poles at A and B.

All the rhumb-lines of one series and all similar lines on spheres concentric with the spherical frame are uniformly and equally extended, all the lines orthogonal to these on the spheres equally compressed.

The third system of orthogonals, viz., radii from the centre of the sphere, are unchanged in length and simply rotated about the axis AB.

The minimum volume of the frame is

$$\frac{2\mathrm{L}}{a} \log \tan\left(\frac{\pi}{4} + \frac{\lambda}{2}\right) \times \left(\frac{1}{\mathrm{P}} + \frac{1}{\mathrm{Q}}\right),$$

where L is the moment of the transmitted couple, $2a = \mathrm{AB}$, and λ is the latitude of small circles about each pole corresponding to the circle of radius r_0 in Case 1.

Melbourne, April 19, 1904.

D.2 Keller's Optimal Column

This reprint is a copy of the paper

"The Shape of the Strongest Column" by Joseph B. Keller, *Archive for Rational Mechanics and Analysis,* 5, 1960, pp. 275–285.

In it Keller solves one of the basic problems of structural optimization. Beyond that, the reader will enjoy the mathematical skills he is so well known for and uses so smoothly here.

This paper is reprinted with the permission of both the author and the publisher, Springer-Verlag.

The Shape of the Strongest Column

JOSEPH B. KELLER

Communicated by C. TRUESDELL

Abstract

The problem of determining that shape of column which has the largest critical buckling load is solved, assuming that the length and volume are given and that each cross section is convex. The strongest column has an equilateral triangle as cross section, and it is tapered along its length, being thickest in the middle and thinnest at its ends. Its buckling load is 61.2% larger than that of a circular cylinder. For columns all of whose cross sections are similar and of prescribed shape—not necessarily convex—the best tapering is found to increase the buckling load by one third over that of a uniform column. This result, which was independently obtained by H. F. WEINBERGER, is originally due to CLAUSEN (1851). For a uniform column, triangularizing is shown to increase the buckling load by 20.9% over that of a circular cylinder. The results lead to isoperimetric inequalities for the buckling loads of arbitrary columns.

1. Introduction

Of all columns having prescribed length and volume and convex cross sections, which is the strongest, *i.e.*, which has the largest critical buckling load? We shall solve this problem and show that the strongest column is not uniform but is tapered, being thickest at its center and thinnest at its ends. We shall also show that the cross section of the strongest column is not a circle but is instead an equilateral triangle. Appropriately tapering a column of circular cross section increases its critical buckling load by one third over that of a uniform column. Changing the cross section from a circle to an equilateral triangle increases the critical buckling load by 20.9%. Thus the combined modifications of tapering and triangularizing increase the critical buckling load by 61.2% over that of a circular cylinder.

Our principal result can also be stated as an isoperimetric inequality. In this form it states that the critical buckling load of any column is less than a given function of the length, volume and YOUNG's modulus of the column. The given function is just the critical buckling load of the strongest column with the same length, volume and YOUNG's modulus. In the course of our analysis two other isoperimetric inequalities arise. One pertains to an untwisted tapered column with similar cross sections and the other to cylindrical columns. The former can also be stated as an isoperimetric inequality for a certain eigenvalue problem, without reference to columns. The latter is related to a geometrical problem which is not completely solved. In the last section we present some results for the nonlinear buckling problem.

For the sake of clarity we first determine the strongest cylindrical column of arbitrary convex cross section. Then we determine the strongest untwisted column, all cross sections of which are similar. Upon combining these two results we find the strongest untwisted column with convex cross sections. Finally we show that this column is also strongest among all columns with convex cross sections. This procedure is also economical since we use the simple results in the final section.

Our analysis is based upon the six equations of equilibrium of a column of arbitrary shape. To these we add expressions for the three stress couples in terms of the curvatures and twist of the column. These three relations are the usual generalizations of the Bernoulli-Euler theory. We then seek solutions of the nine equations subject to the conditions that the stress couples vanish at the ends of the column and that the applied load have a prescribed value. To determine the critical buckling load we linearize the equations about the straight unbuckled state. The linearized problem is an eigenvalue problem and has non-zero solutions only for special values of the load. The smallest of these is the critical buckling load. It depends upon the shape of the column. We next apply variational techniques to determine that shape which maximizes the critical buckling load. In this way we obtain the results described above.

I wish to thank CLIFFORD TRUESDELL for proposing this problem to me. He also suggested it to HANS F. WEINBERGER, who independently obtained the results of Section 3 on untwisted columns with circular cross sections. However, as TRUESDELL later pointed out, that result had been obtained by CLAUSEN[1] in 1851. LAGRANGE[2] in 1773 first treated that problem but arrived at the wrong result due to computational errors[3].

2. Cylindrical columns of arbitrary convex cross section

For simplicity let us begin by considering cylindrical columns of arbitrary cross section. Let L and V denote the length and volume of such a column and $A = V/L$ its cross sectional area. Let the x-axis pass through the centroid of each cross section of the column in its straight or unbuckled state. Suppose that I' is the largest moment of inertia of the cross section about a line in its

[1] Über die Form architektonischer Säulen. Bulletin physico-mathématique de l'Académie 9, 368–379 (St. Pétersburg 1851). Also Mélanges Mathématiques et Astronomiques, Tome I (1849–1853), pp. 279–294. St. Petersburg 1853. This work is summarized in TODHUNTER, I., & K. PEARSON, A History of the Theory of Elasticity and of the Strength of Materials, vol. 2, pp. 325–329. Cambridge 1893.

[2] Sur la figure des colonnes. Miscellanea Taurinensia (Royal Society of Turin) Tomus V, p. 123, 1770–1773. Also Oeuvres, vol. 2, pp. 125–170. Gauthier-Villars 1868. Summarized in TODHUNTER & PEARSON, vol. 1, pp. 66–67. Cambridge 1886. A more complete early history of this and related problems is given by TRUESDELL, The Rational Mechanics of Flexible or Elastic Bodies, 1638–1788, L. Euleri Opera Omnia II 11_2 (in press).

[3] These errors are pointed out in his Oeuvres by the editor. They are not mentioned in TODHUNTER & PEARSON.

plane through the centroid and that the line is the y-axis. Similarly let the minimum moment of inertia be I. Then the corresponding line is the z-axis.

Let us consider a state of equilibrium of the column in which a compressive load T_0 is applied longitudinally at its ends. Let us further suppose that the deflection of the column from the straight unbuckled state is small. Then the equilibrium is governed by linear equations and every state is a superposition of a deflection in the xy-plane and another in the xz-plane. These deflections $y(x)$ and $z(x)$ satisfy the equations

$$y_{xx} + \frac{T_0}{EI} y = 0, \quad y(0) = y(L) = 0, \tag{1}$$

$$z_{xx} + \frac{T_0}{EI'} z = 0, \quad z(0) = z(L) = 0. \tag{2}$$

The boundary conditions in (1) and (2) apply if the column is pinned at its ends, and E denotes the YOUNG's modulus of the column material.

The least positive value of T_0 for which (1) has a non-trivial solution is $\pi^2 E I L^{-2}$ while for (2) it is $\pi^2 E I' L^{-2}$. The critical buckling load is the minimum of these two values. Since by assumption $I \leqq I'$, we have for the critical load

$$T_0 = \pi^2 E I L^{-2}. \tag{3}$$

The strongest column is that for which T_0 is as large as possible. Since E, L and V, and therefore A, are prescribed, the strongest column is one for which I is largest. But I can be made arbitrarily large if the cross section is nonconvex. Therefore the problem of finding a strongest column makes sense only if the cross section is required to be convex. Then the cross section of the strongest column is the solution of the following geometrical problem. Problem: Of all plane convex domains of area A, which has the largest value of I? Here I denotes the minimum value of the moment of inertia of the domain about any line through its centroid.

This problem does not seem to have been investigated previously. It is easy to show that the domain which solves the problem has the same moment of inertia about every line through its centroid which lies in the plane. However this domain is not a circle. This follows from the facts that the circle has the smallest polar moment of inertia of all plane figures of given area and that I is one half the polar moment for any figure having the same moment I about every line through the centroid. Every regular polygon has the same moment about every line, through the centroid, and of these the equilateral triangle has the largest I for given area. Therefore it has been conjectured by PETER UNGAR that the equilateral triangle is the solution of the above problem. GEORGE STELL has proved that it is the solution among all figures having an n-fold symmetry axis with $n \geqq 3$. The complete solution has not yet been given, but there is little doubt that it is the equilateral triangle. If this is the case, we have the isoperimetric inequality for convex domains

$$I \leqq \frac{A^2}{6\sqrt{3}} = A^2(.096225). \tag{4}$$

Equality holds for the equilateral triangle.

By using (4), (3) yields the isoperimetric inequality for the critical buckling load of cylindrical columns with convex cross sections

$$T_0 \leqq \frac{\pi^2 E A^2}{6\sqrt{3}\, L^2} = \frac{E A^2}{L^2} (.949703). \tag{5}$$

Equality holds for the equilateral triangular cross section. The numerical coefficient in (4) is 20.9% larger than the value $\pi/4 = .785$ for the circular cross section.

3. Tapered untwisted columns with similar cross sections

Let us now consider tapered columns. We shall still assume that all cross sections are similar and that the principal axes of inertia are parallel. Thus the column is not twisted. However the area $A(x)$ of the cross section may vary with x subject to the condition

$$\int_0^L A(x)\, dx = V. \tag{1}$$

The moments I and I' are related to A by the equations

$$I(x) = \alpha A^2(x), \qquad I'(x) = \alpha' A^2(x). \tag{2}$$

The constants α and α', with $\alpha \leqq \alpha'$, depend upon the shape of the cross section. Now (1) becomes

$$y_{xx} + \frac{T_0}{E \alpha A^2(x)}\, y = 0, \qquad y(0) = y(L) = 0. \tag{3}$$

We seek that function $A(x)$ satisfying (1), which maximizes the lowest eigenvalue T_0 of (3). This function will be the area variation of the strongest column.

Before solving this problem we first introduce the new variable $\xi = x/L$ and the eigenvalue parameter $\lambda = T_0 L^2/\alpha E$ and consider y and A as functions of ξ. Then (3) and (1) become

$$y'' + \lambda A^{-2}(\xi)\, y = 0, \qquad y(0) = y(1) = 0, \tag{4}$$

$$\int_0^1 A(\xi)\, d\xi = V/L. \tag{5}$$

Now we assume that $A(\xi)$ is the solution. We introduce a family of functions $A(\xi, \varepsilon)$ depending differentiably on a parameter ε, satisfying (5), and containing the solution $A(\xi) = A(\xi, 0)$. Then for each ε, (4) will have a solution $y(\xi, \varepsilon)$ with the corresponding lowest eigenvalue $\lambda(\varepsilon)$, both of which depend differentiably on ε. Next we differentiate (4) and (5) with respect to ε, obtaining

$$y_\varepsilon'' + \lambda A^{-2} y_\varepsilon = 2\lambda A_\varepsilon A^{-3} y - \lambda_\varepsilon A^{-2} y, \qquad y_\varepsilon(0) = y_\varepsilon(1) = 0, \tag{6}$$

$$\int_0^1 A_\varepsilon\, d\xi = 0. \tag{7}$$

The derivative y_ε satisfies an inhomogeneous linear differential problem (6) and y satisfies the corresponding homogeneous problem (4). Therefore the right side of the differential equation in (6) must be orthogonal to y. At $\varepsilon = 0$, $\lambda_\varepsilon = 0$ since by assumption λ is a maximum at $\varepsilon = 0$. The orthogonality condition is then

$$\int_0^1 A_\varepsilon (A^{-3} y^2)\, d\xi = 0. \tag{8}$$

Since (8) must hold for every $A_\varepsilon(\xi)$ satisfying (7), it follows that $A^{-3}y^2$ is a constant which we denote by $c^{\frac{3}{2}}$. Thus for the strongest column we have

$$y^2 = c^{\frac{3}{2}} A^3. \tag{9}$$

To find y we first eliminate A from (4) by means of (9) and obtain

$$y'' + \lambda c\, y^{-\frac{1}{3}} = 0, \qquad y(0) = y(1) = 0. \tag{10}$$

The solution of (10) is

$$\left|\xi - \frac{1}{2}\right| = \frac{1}{2} + \frac{1}{\pi}\left(\frac{y}{y_0}\right)^{\frac{1}{3}}\left[1 - \left(\frac{y}{y_0}\right)^{\frac{2}{3}}\right]^{\frac{1}{2}} - \frac{1}{\pi}\sin^{-1}\left(\frac{y}{y_0}\right)^{\frac{1}{3}}. \tag{11}$$

Here y_0 is given by

$$y_0^{\frac{4}{3}} = \frac{2}{3\pi}(\lambda c)^{\frac{1}{2}}. \tag{12}$$

Upon replacing y in (11) by A, with the aid of (9), we obtain the following equation for the solution $A(\xi)$:

$$\left|\xi - \frac{1}{2}\right| = \frac{1}{2} + \frac{1}{\pi}\left[\frac{A}{A_0} - \left(\frac{A}{A_0}\right)^2\right]^{\frac{1}{2}} - \frac{1}{\pi}\sin^{-1}\left(\frac{A}{A_0}\right)^{\frac{1}{2}}. \tag{13}$$

The constant $A_0 = A(\frac{1}{2})$ in (13) can be determined as a function of V/L by inserting (13) into (5). This yields

$$V/L = 2\int_0^{\frac{1}{2}} A(\xi)\, d\xi = 2\int_0^{A_0} \xi\, dA = \frac{3A_0}{4}. \tag{14}$$

To find λ let us use (9) to eliminate y from (4), which becomes

$$A A'' + \tfrac{1}{2}(A')^2 + \tfrac{2}{3}\lambda = 0. \tag{15}$$

From (13) we find that $A'(\frac{1}{2}) = 0$ and $A''(\frac{1}{2}) = \pi^2 A_0/2$. Upon inserting these values into (15) and using (14) for A_0 we obtain

$$\lambda = \frac{4\pi^2}{3}\left(\frac{V}{L}\right)^2. \tag{16}$$

Now from the definition of λ we find the critical buckling load of the strongest column. The result may be stated as an isoperimetric inequality

$$T_0 \leq \frac{4\pi^2}{3}\alpha V^2 L^{-4} E. \tag{17}$$

Equality holds only for the strongest column of given volume, the shape of which is given by (13). For it the critical buckling load is $\frac{4}{3}$ times as large as that of a cylindrical column of the same volume, length and cross section.

Let us now combine the results of this and the preceding section. Then we find that the strongest untwisted column, all of whose cross sections are similar and convex, is tapered according to (13) and has an equilateral triangular cross section. Upon using the value $\alpha = (6\sqrt{3})^{-1}$ appropriate to this triangle, we obtain from (17) the isoperimetric inequality

$$T_0 \leq \frac{2\pi^2}{9\sqrt{3}}\frac{V^2 E}{L^4} = \frac{V^2 E}{L^4}(1.26627). \tag{18}$$

Equality holds only for the column just described. For it the critical buckling load is 1.612 times as large as for a circular cylindrical column. In the remaining sections we shall show that (18) also holds for twisted columns, without the restriction that the cross sections be similar.

Our result (16) can be expressed as an isoperimetric inequality for the eigenvalue problem (4) without reference to columns. It can be formulated thus: Let λ be the lowest eigenvalue of the problem

$$y'' + \lambda q(x)\, y = 0, \quad y(0) = y(L) = 0, \quad q(x) > 0. \tag{19}$$

Then

$$\lambda \leq \frac{4\pi^2}{3L^2}\left[\int_0^L q^{-\frac{1}{2}}(x)\, dx\right]^2. \tag{20}$$

Equality holds if $q(x) = A^2(x/L)$, where A is given by (13).

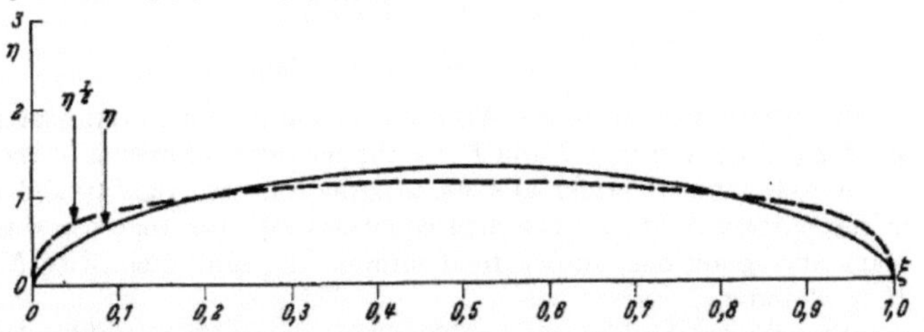

Fig. 1. The area (solid curve) and square root of the area (dashed curve) of the strongest column as functions of distance along the column. These curves are based upon equations (21) and (22). The vertical coordinate is $\eta = A\,L/V$ or $\eta^{\frac{1}{2}} = (AL/V)^{\frac{1}{2}}$, and the horizontal coordinate is $\xi = x/L$. Here $A(x)$ is the cross sectional area at the distance x along the column, L is the length and V the volume of the column

The equation (13) for the shape of the strongest column can be simply expressed in terms of a parameter ϑ which varies from 0 to π. Then, as CLIFFORD TRUESDELL has observed, (13) and (14) yield

$$A = \frac{4V}{3L}\sin^2\vartheta, \tag{21}$$

$$\xi = \frac{1}{\pi}\left(\vartheta - \frac{1}{2}\sin 2\vartheta\right). \tag{22}$$

4. Arbitrary columns with convex cross sections

Let us now consider any column of length L and volume V, all cross sections of which are convex. Among them we seek the strongest one. To find it we begin by considering any such column in a state of equilibrium in the absence of external distributed forces or couples. The vanishing of the net force and torque on each portion of the column leads to six equations [LOVE p. 387 eq. 10, p. 388 eq. 11][1]

$$N_s - N'\tau + T\varkappa' = 0, \tag{1}$$

$$N_s' - T\varkappa + N\tau = 0, \tag{2}$$

$$T_s - N\varkappa' + N'\varkappa = 0, \tag{3}$$

$$G_s - G'\tau + H\varkappa' - N' = 0, \tag{4}$$

$$G_s' - H\varkappa + G\tau + N = 0, \tag{5}$$

$$H_s - G\varkappa' + G'\varkappa = 0. \tag{6}$$

[1] LOVE, A. E. H.: A Treatise on the Mathematical Theory of Elasticity, 4th ed. Cambridge Univ. Press. 1927.

In these equations N and N' are the shearing forces, T is the tension, G and G' are the flexural couples, H is the torsional couple, $\varkappa$ and $\varkappa'$ are the components of curvature of the central line and τ is the twist of the rod. The subscript s denotes differentiation with respect to arclength along the central line. The forces and couples are expressed as components along the principal torsion-flexure axes of the rod.

In the usual approximate theory of a naturally straight thin rod, which is a generalization of the Bernoulli-Euler theory, it is assumed that the stress couples are related to the curvatures and twist of the rod by the equations [LOVE p. 388, eq. 12, p. 397, eq. 28][1]

$$G = E I \varkappa, \tag{7}$$

$$G' = E I' \varkappa', \tag{8}$$

$$H = C(\tau - \tau_0). \tag{9}$$

In these equations E is the YOUNG's modulus of the rod material, which is assumed to be isotropic, I and I' are the moments of inertia of the cross section of the rod about its principal axes, C is the torsional rigidity and τ_0 is the twist in the unstressed state. The nine equations (1)—(9), together with appropriate boundary conditions, suffice to determine the nine quantities N, N', T, G, G', $H, \varkappa, \varkappa'$ and τ.

We wish to consider the buckling of a naturally straight column subjected to a compressive axial load. In the straight or unbuckled state $\varkappa = \varkappa' = 0$. Then from (7) and (8) it follows that $G = G' = 0$. But then (4) and (5) yield $N = N' = 0$. Therefore (3) and (4) become $T_S = H_S = 0$ or $T = -T_0 =$ constant, $H =$ constant. We shall assume that no twisting couple is applied at the ends of the rod. Consequently $H = 0$, and then (9) shows that $\tau = \tau_0$. Thus in the straight state only T and τ may differ from zero. If we consider a buckled state which is sufficiently close to the unbuckled state, we may assume that all quantities except T and τ are small while T and τ are close to their values in the unbuckled state. Then we may linearize (1)—(9), which become

$$N_S - N'\tau_0 - T_0\varkappa' = 0, \tag{10}$$

$$N'_S + T_0\varkappa + N\tau_0 = 0, \tag{11}$$

$$T_s = 0, \tag{12}$$

$$G_s - G'\tau_0 - N' = 0, \tag{13}$$

$$G'_s + G\tau_0 + N = 0, \tag{14}$$

$$H_s = 0. \tag{15}$$

From (15) and the boundary condition that no couples act at the ends, it follows that $H = 0$, and (9) still shows that $\tau = \tau_0$. Similarly (12) yields $T =$ constant, and if T_0 is the applied load, then $T = -T_0$.

The linearized buckling problem is that of solving (10), (11), (13), (14), (7) and (8) for $N, N', G, G', \varkappa$ and $\varkappa'$ when I, I' and τ_0 are given functions of s, and E is a given constant. At the endpoints $s = 0$ and $s = L$ we require that

[1] See the previous footnote.

no couples act. Thus

$$G(0) = G'(0) = G(L) = G'(L) = 0. \tag{16}$$

This is an eigenvalue problem and will have non trivial solutions only for certain special values of the constant T_0. The smallest buckling load is the least value of T_0 for which a non trivial solution exists.

If we eliminate $\varkappa$ and $\varkappa'$ from (10) and (11) by means of (7) and (8), we obtain

$$N_s - N'\tau_0 - \frac{T_0}{EI'}G' = 0, \tag{17}$$

$$N_s' + N\tau_0 + \frac{T_0}{EI}G = 0. \tag{18}$$

We must now solve (13), (14), (17) and (18) for N, N', G and G' subject to (16), which is possible only if T_0 is an eigenvalue. We now seek that column, characterized by τ_0, I and I', for which the first eigenvalue is largest. It must have the volume V, so

$$\int_0^L A(s)\, ds = V. \tag{19}$$

5. The variational problem

Let us now consider a one parameter family of columns depending differentiably upon the parameter ε and containing the strongest column at $\varepsilon = 0$. Thus τ_0, I and I' are all differentiable functions of ε as well as of s. Then the solution of the buckling problem will also depend differentiably on ε. Then we obtain from (13), (14), (17), (18), (16) and (19) of Section 4 respectively, with $\partial/\partial\varepsilon$ denoted by a dot, the variational equations

$$\dot{G}_s - \dot{G}'\tau_0 - \dot{N}' = G'\dot{\tau}_0, \tag{1}$$

$$\dot{G}_s' + \dot{G}\tau_0 + \dot{N} = -G\dot{\tau}_0, \tag{2}$$

$$\dot{N}_s' + \dot{N}\tau_0 - \frac{T_0}{EI}\dot{G} = -N\dot{\tau}_0 - \left(\frac{T_0}{EI}\right)^{\bullet} G, \tag{3}$$

$$\dot{N}_s - \dot{N}'\tau_0 + \frac{T_0}{EI'}\dot{G}' = N'\dot{\tau}_0 + \left(\frac{T_0}{EI'}\right)^{\bullet} G', \tag{4}$$

$$\dot{G}(0) = \dot{G}'(0) = \dot{G}(L) = \dot{G}'(L) = 0, \tag{5}$$

$$\int_0^L \dot{A}\, ds = 0. \tag{6}$$

The equations (1)—(5) form an inhomogeneous system for the four derivatives $\dot{G}$, $\dot{G}'$, $\dot{N}$ and $\dot{N}'$. The corresponding homogeneous system is just that satisfied by G, G', N, and N', *i.e.*, (13), (14), (17), (18) and (16) of Section 4. Therefore in order that the inhomogeneous system have a solution, the right sides of (1) to (4) must be orthogonal to all the solutions of the adjoint homogeneous system. The only solution of this system is $(-N', N, -G', G)$. Thus the orthogonality condition is

$$\int_0^L \left[-2(N'G' + NG)\dot{\tau}_0 - G'^2\left(\frac{T_0}{EI'}\right)^{\bullet} - G^2\left(\frac{T_0}{EI}\right)^{\bullet}\right] ds = 0. \tag{7}$$

In (7) we choose $\dot{I}'=\dot{I}=0$ and assume that $\dot{T}_0=0$ at $\varepsilon=0$. Then since (7) must hold for every choice of $\dot{\tau}_0$, we conclude that for $\varepsilon=0$

$$N'G' = -NG. \tag{8}$$

Let us now deduce some conclusions from the foregoing equations. First we multiply (13) of Section 4 by $-G'$ and (14) of Section 4 by G, add and use (8) to obtain

$$G'_s G - G' G_s + (G'^2 + G^2)\,\tau_0 = 0. \tag{9}$$

We rewrite (9) and define β_s as follows:

$$\frac{(G/G')_s}{(G'/G)^2+1} = -\tau_0 \equiv -\beta_s. \tag{10}$$

Upon integrating (10) and adding an appropriate constant to β, we have

$$G'/G = -\tan\beta. \tag{11}$$

Let us define g by

$$g^2 = G'^2 + G^2. \tag{12}$$

Then from (11) and (12) we find

$$G' = -g\sin\beta, \qquad G = g\cos\beta. \tag{13}$$

Now we define n by

$$n^2 = N'^2 + N^2. \tag{14}$$

Then from (8) and (14) we get

$$N' = n\cos\beta, \qquad N = n\sin\beta. \tag{15}$$

Now we multiply (13) of Section 4 by G, (14) of Section 4 by G' and add, obtaining

$$\tfrac{1}{2}\,(G^2 + G'^2)_s - GN' + G'N = 0. \tag{16}$$

By using (13) and (15) in (16) we find

$$g\,g_s - n\,g = 0. \tag{17}$$

Hence, if $g \neq 0$, we have

$$g_s = n. \tag{18}$$

We now insert (13), (15) and (18) into (17) and (18) of Section 4 and obtain

$$\left(g_{ss} + \frac{T_0}{EI'}\,g\right) g\sin\beta = 0, \tag{19}$$

$$\left(g_{ss} + \frac{T_0}{EI}\,g\right) g\cos\beta = 0. \tag{20}$$

If $g\sin\beta \neq 0$ and $g\cos\beta \neq 0$, we conclude from (19) and (20) that

$$I = I', \tag{21}$$

$$g_{ss} + \frac{T_0}{EI}\,g = 0. \tag{22}$$

From (21) and (2) of Section 2, with $\alpha=\alpha(s)$, $\alpha'=\alpha'(s)$, it follows that $\alpha=\alpha'$. If we set $\lambda = T_0/\alpha E$, we may rewrite (22) in the form

$$g_{ss} + \lambda A^{-2} g = 0. \tag{23}$$

From (16) of Section 4 and the definition of g, we have

$$g(0) = g(L) = 0. \tag{24}$$

If λ denotes the lowest eigenvalue of the problem (23), (24), then $T_0 = \alpha E \lambda$. Thus for a fixed A^2, T_0 is maximized by maximizing α. Since $\alpha = \alpha'$, the problem of maximizing α is just the same geometrical problem considered in Section 2. Then since the best value of α is a constant, independent of s, the problem of finding $A(s)$ to maximize λ is just that solved in Section 3. Thus if $\sin\beta \neq 0$ and $\cos\beta \neq 0$, the strongest column is that described toward the end of Section 3.

The exceptional case in which $\sin\beta = 0$ or $\cos\beta = 0$ occurs if $\tau_0 = 0$, *i.e.*, if the column is untwisted. Then we must compare the lowest eigenvalues for the two cases $\beta = 0$ and $\pi/2$. The smaller of the two is the critical buckling load. The former case yields (22)—(24), and the latter yields the same equation with α' in place of α. Thus the critical buckling load is the smaller of $\alpha E \lambda$ and $\alpha' E \lambda$. The determination of the best cross section is thus just that of solving the geometrical problem of Section 2, and then the determination of the best $A(s)$ is the problem of Section 3. Thus the strongest column is that found in Section 3.

6. The nonlinear buckling problem

In the preceding sections we considered the linear buckling problem. That problem was obtained by linearizing the original nonlinear problem around the straight unbuckled state. It yields the surprising result that there is no buckled state unless the load is equal to an eigenvalue. But we expect buckled states to exist for any load exceeding the lowest eigenvalue. In fact, if we consider deflections in a single plane, we expect n buckled states to exist when the load exceeds the nth eigenvalue of the linear problem, but does not exceed the $(n+1)^{\text{st}}$. These buckled states correspond to the n buckling modes associated with the n eigenvalues less than the load. This expectation is borne out for cylindrical columns for which the explicit solutions of the nonlinear problem (the elastica problem) are known. However for tapered columns, such as those considered here, it has not yet been shown that this is the case. The corresponding statement has been proved to be true by I. I. KOLODNER [1] for a somewhat similar problem.

As a first step toward verifying this conjecture, we shall show that there is a buckled state for any load slightly larger than any buckling load. To do so we shall consider deflections of an untwisted column in a single plane and let $\vartheta(s)$ denote the angle between the column and a fixed axis in this plane. Then, as follows from the equations (1)—(9) and (16) of Section 4, $\vartheta(s)$ satisfies the equations

$$(I\vartheta_s)_s + \lambda \sin\vartheta = 0, \quad \vartheta_s(0) = \vartheta_s(L) = 0. \tag{1}$$

The constant $\lambda = T_0 E^{-1}$ and the positive function $I(s)$ are given, and $\vartheta(s)$ is to be found.

It is convenient to introduce the constant a and the function u defined by

$$\vartheta = a u, \quad a = \vartheta(0). \tag{2}$$

---

[1] Comm. Pure Appl. Math. 8, 395—408 (1955).

Then u satisfies the equations

$$(I u_s)_s + \lambda a^{-1} \sin(a u) = 0, \quad u(0) = 1, \quad u_s(0) = 0, \quad u_s(L) = 0. \tag{3}$$

If we let a tend to zero in (3), we obtain the linear problem

$$(I u_s)_s + \lambda u = 0, \quad u(0) = 1, \quad u_s(0) = 0, \quad u_s(L) = 0. \tag{4}$$

This problem has a discrete set of increasing positive eigenvalues λ_n^0 and eigenfunctions u_n^0.

Let us suppose that a is given and that u and λ are both to be found. Then we assume that for a small

$$u = u_n^0 + a^2 u_n^1 + \cdots, \tag{5}$$

$$\lambda = \lambda_n^0 + a^2 \lambda_n^1 + \cdots. \tag{6}$$

We have expanded in powers of a^2 because (3) is even in a. Insertion of (5) and (6) into (3) and a simple calculation yield the result

$$\lambda_n^1 = \frac{\lambda_n^0}{6} \frac{\int_0^L (u_n^0)^4 \, ds}{\int_0^L (u_n^0)^2 \, ds}. \tag{7}$$

Since $\lambda_n^1 > 0$, we can solve (6) for a in terms of λ provided $\lambda \geqq \lambda_n^0$. If we denote the solution by $a_n(\lambda)$, we have

$$a_n(\lambda) = \left(\frac{\lambda - \lambda_n^0}{\lambda_n^1}\right)^{\frac{1}{2}} [1 + O(\lambda - \lambda_n^0)]. \tag{8}$$

By using this result, it is easy to prove that (3), and therefore (1), has a solution for $\lambda - \lambda_n^0$ sufficiently small and non-negative. This shows that the nonlinear problem has buckled solutions for any load in a certain half-neighborhood of each buckling load of the linear problem. It also shows that the lowest buckling load of the linear problem is also the lowest buckling load of the nonlinear problem, at least for slightly buckled states. Of course the conjecture concerning the number of buckled states remains to be proved.

The research reported in this paper has been sponsored by the Office of Naval Research under Contract No. (285) 46.

Institute of Mathematical Sciences
New York University

(Received March 4, 1960)

D.3 The Paper of Venkayya, Khot, and Berke

This paper is a reprint of

> "Application of Optimality Criteria Approaches to Automated Design of Large Practical Structures" by V. B. Venkayya, N. S. Khot, and L. Berke, *Conference Proceedings No. 123,Second Symposium on Structural Optimization,* AGARD-CP-123, Milan, Italy, April 1973, pp. 3.1–3.19.

The reader will by now have encountered the so-called optimality criteria methods of structural optimization. What may not be clear so far is the extent to which these methods dominated the early years of structural optimization. Beyond that fact, the reader will enjoy the breadth of this work which is quite remarkable.

APPLICATION OF OPTIMALITY CRITERIA APPROACHES TO AUTOMATED DESIGN OF LARGE PRACTICAL STRUCTURES

By

V. B. Venkayya
Aerospace Engineer

N. S. Khot
Aerospace Engineer

and

L. Berke
Technical Manager

Synthesis Group
Air Force Flight Dynamics Laboratory (FBR)
Wright-Patterson Air Force Base, Ohio 45433

SUMMARY

A unified approach for the derivation of optimality criteria for the design of optimum structures is presented. The design conditions included requirements of generalized stiffness, specified displacements, dynamic stiffness and stiffness for general stability. Recursion relations for achieving the optimality criteria are derived for all these cases. Both single and multiple stiffness requirements are considered. The formulation considers isotropic, anisotropic and layered composite structures.

Four categories of design examples are selected to illustrate the versatility of the method. The first group is designed for static loads with constraints on stresses, displacements and minimum sizes. The second group of examples are designed for dynamic loads with periodic time dependence. The third group consists of layered composite structures subjected to static loads. The fourth group are with static stability requirements. The results of many of these examples are compared with those available in the literature.

1. INTRODUCTION

The current trend in aircraft design technology is the computer automation of as much of the design process as feasible. During the 1960's all of the disciplines participating in the design activity have developed extensive computer capabilities, however, with little thought to interdisciplinary data communication. Considerable pioneering work has already been done in the early 1970's by industry and government to integrate these capabilities into automated design procedures. It is safe to say that the remainder of the 1970's will be marked by intensive development work in this area.

The formal optimization techniques employed in these integrated design procedures, both within the various disciplines and for overall design optimization, have to be extremely efficient. Structural analysis coupled with optimization both at the local or structural component and at the structural assembly level forms an important module of any integrated aircraft design capability and with a potential to consume large computer resources.

The automated optimum sizing of large complex structural assemblies, with perhaps a few thousand size variables at the final design level, is a necessary ingredient of any such procedure because the laborious and time consuming manual sizing cannot be tolerated in an otherwise automated approach created for quick turnaround. This paper describes the in-house and some of the contracted development work performed in the Structures Division of the Flight Dynamics Laboratory in the specific area of automated optimum sizing for complex structural systems simulated by large finite element assemblies.

Structural optimization is an order of magnitude more difficult a task than. analysis and this is reflected by the large difference in their relative rate of development. Until the late 1950's the method of simultaneous failure modes for structural components and fully stressed design for structural systems, two essentially intuitive optimality criteria approaches, formed the backbone of structural optimization practice. At the end of the 1950's Klein, Schmit and others proposed a rational approach to structural optimization by formulating it as a problem in mathematical programming. This new approach relied on direct numerical search techniques to find the optimum of a function of several variables subject to equality and inequality constraints. The generality of the formulation made this approach extremely well suited to optimize structures with a mixture of geometric and size variables and with stress, deflection, stability and other constraints. The initial developments not only proved feasibility but, as a side product, also pointed out the shortcomings of the two time honored optimization methods mentioned earlier, somewhat discrediting them in the process.

The developments in structural optimization based on numerical search techniques mushroomed quickly because it suddenly opened up a new promising territory for research work with computer applications attracting the interest of many capable researchers to a territory formerly almost the exclusive domain of the structural engineer. In the wake of the demonstrated successes, and anticipating further improvements in numerical search techniques, computer capability and structural analysis methods, an effort was initiated to investigate the feasibility of applying mathematical programming techniques to the optimization of large structural systems using finite element methods as the analysis module. It was hoped not only to replace fully stressed design methods, applicable to strength design only, but to couple the general purpose optimization capability of direct numerical search techniques with the general purpose analysis capability of the finite element methods to handle structural optimization problems with a wide range of behavior and side constraints.

As described [1] the effort resulted in mixed success. It demonstrated technical feasibility and rapidly increasing economical unfeasibility for problems with more than a few dozen design variables.

By the middle of the 1960's, as a result of these and other developments, a certain disillusionment set in with numerical search procedures for large structural systems, while applications to a wide range of problem, all characterized by a relatively small number of variables, showed steady improvements. It was becoming clear that for the special problem of the simple but large number of size variables, the generality of the mathematical programming approach extracted an unacceptable toll, and a special approach was needed that capitalized on the special nature of the problem by making use of structural theorems and of special properties of an optimum, structure, if optimization capability was not to trail far behind analysis capabilities.

The first systematic research effort towards such a capability was initiated by Venkayya in 1968 as an in-house investigation and resulted in an extremely efficient method for stress and displacement constrained problems. It was first reported in [2] and a detailed study with some modifications was presented in [3]. For strength design it uses a simple recursion formula, based on an optimality criterion statement, to drive the design towards an optimum. For displacement constraints an efficient numerical search scheme based on constraint gradients was developed resulting in what is basically a combined method. A recursion relation based on the optimality criteria for displacement constraint problems was derived in [4]. This recursion relation eliminated the need for determining constraint gradients which are computationally expensive.

In 1969 an effort to develop a computer program for the optimization of large aircraft components was initiated. The result of this effort was the Automated Structural Optimization Program (ASOP), [5]. This program uses a combined approach based on stress ratio design and the numerical search similar to the method presented in [2]. At present this program is being used extensively by the Air Force Flight Dynamics Laboratory for optimization of wing and stabilizer structures.

To further explore the possibilities of improving the optimality criteria approaches, two small efforts were initiated. The first effort was to establish rigorous optimality criteria for finite element systems with multiple constraints and loading conditions. The results of this investigation were reported in [6]. The purpose of the second effort was to study the convergence characteristics of optimality criteria approaches and to evolve a working method for the displacement constrained problems. The results of this study were reported in [7].

An optimality criterion for the design of structures subjected to dynamic loads was presented in 1971 [8]. The iterative procedure based on this criterion is called design in the dynamic mode. The dynamic mode may consist of a single natural mode or a linear combination of a set of natural modes of the structure depending on the nature of the forcing function.

An optimality criterion and an iterative algorithm for the design of fiber reinforced layered composite structures was derived in 1971 [9]. This method was successfully used for the design of plate and wing structures. The optimality criteria approach was extended to include static stability constraints in [10] and [11].

It is evident from the foregoing discussion that most structural optimization problems of practical interest can be handled very efficiently by optimality criteria approaches. This paper attempts to highlight some of the merits and deficiencies of optimality criteria approaches when applied to practical aerospace structures.

II. Optimality Criteria for Minimum Weight Structures

The purpose of this section is to present a unified approach for the derivation of optimality criteria for various design conditions. The objective in all cases is to minimize the weight of the structure while satisfying the desired stiffness characteristics under specified loading conditions. The configuration of the structure is assumed to be fixed and the sizes of the elements are the only design variables. The total structure is discretized into "m" finite elements. Under these conditions the total weight of the structure may be expressed as

$$W = \sum_{1}^{m} \rho_i A_i l_i \qquad (1)$$

where ρ_i is the mass density and the product $A_i\ l_i$ represents the volume of the element.

If the structure is subjected to an external load vector E, the load displacement relations are given by

$$\underline{R} = \underline{K}\ \underline{r} \qquad (2)$$

where r is the displacement vector and $\underline{K}$ is the generalized stiffness of the structure. The structure and its element stiffness matrices are related by

$$\underset{\sim}{K} = \sum_{i=1}^{m} \underset{\sim}{a}_i^t\ \underset{\sim}{k}_i\ \underset{\sim}{a}_i \qquad (3)$$

The matrix a, is the compatibility matrix and its elements can be obtained by kinematic reasoning alone. The ith element stiffness matrix depends only on the ith design variable. The corresponding load displacement relations for a dynamically loaded structure are

$$\underset{\sim}{M}\ \ddot{\underset{\sim}{r}} + \underset{\sim}{K}\ \underset{\sim}{r} = \underset{\sim}{R}(t) \qquad (4)$$

where M is the generalized mass matrix of the structure. In case of consistent mass formulation the element and structure mass matrices are related by an expression similar to Equation (3). The (•) represents a time derivative. $\underline{R(t)}$ is the time dependent forcing function.

For the case of free vibration Equation (4) may be written as

$$\omega^2\ \underset{\sim}{M}\ \underset{\sim}{r} = \underset{\sim}{K}\ \underset{\sim}{r} \qquad (5)$$

where ω represents the natural frequency of the structure. The Raleigh quotient for frequency of vibration is given by

$$\omega_j^2 = \frac{X_j^t \, K \, X_j}{X_j^t \, M \, X_j} \tag{6}$$

where $\underline{X}_j$ is the jth natural mode of vibration. The static stability of the structure is governed by

$$[K + \omega K_G] \, r = 0 \tag{7}$$

where K_G is the geometric stiffness matrix of the structure. The Rayleigh quotient for the buckling load is given by

$$\omega_j = \frac{X_j^t \, K \, X_j}{X_j^t \, K_G \, X_j} \tag{8}$$

where X_j is the jth buckled mode and ω_j is the critical load parameter.

If the desired stiffness requirement is denoted by the quantity Z, a new function φ consisting of the weight and the stiffness requirement can be written as

$$\varphi = \sum_1^m \rho_i A_i l_i \pm \frac{1}{\lambda} Z \tag{9}$$

where I/λ is the undetermined Lagrangian multiplier. Minimization of of φ with respect to the design variables gives the condition for the stationary value of the weight with the constraint condition Z.

$$\frac{\partial \phi}{\partial A_i} = \rho_i l_i \pm \frac{1}{\lambda} \frac{\partial}{\partial A_i} Z = 0 \tag{10}$$

Now four stiffness requirements are considered in evaluating Equation (10).

Case 1: Minimum Weight Structure for Generalized Stiffness

If $\underline{R}$ and $\underline{r}$ are the generalized force and the corresponding displacement vectors, the scalar product $\underline{R}^t\underline{r}$ is an inverse measure of the stiffness of the structure. If Z is denoted by

$$Z = \frac{1}{2} \underset{\sim}{R}^t \underset{\sim}{r} \qquad (11)$$

the object is to find the optimality condition for the minimum weight structure subject to the stiffness requirement Z.

Equations (2), (3), (11) and the minimum condition Equation (10) give the Lagrangian multiplier in the form

$$\lambda = e_i \qquad (12)$$

where e_i is the ratio of the average strain energy density to the mass density of the ith element and is given by

$$e_i = \frac{1}{2} \frac{\underset{\sim}{r}^t \underset{\sim}{K}_i \underset{\sim}{r}}{\rho_i v_i} \qquad (13)$$

where $\underline{K}_i$ is the element stiffness matrix of the ith element in general ized coordinates and v_i is its volume.

It is evident from Equation (12) that the minimum condition (Equation 10) can be satisfied only when the ratio of the average strain energy density to the mass density is the same for all the elements. Now the statement of the optimality criterion for the general ized stiffness requirement is as follows:

"The optimum structure for a generalized stiffness requirement is the one in which the ratio of the average strain energy density to its mass density is the same for all its elements."

The above optimality criterion is the same as the one given material.

When stiffness requirements are more than one, as in the in [2] and [3], when the structure is built from the same case of multiple loading conditions and stiffness requirements, the function φ can be written as

$$\phi = \sum_{i=1}^{m} \rho_i A_i l_i + \frac{1}{\lambda_1} Z_1 + \cdots + \frac{1}{\lambda_p} Z_p \tag{14}$$

where $Z_1 \ldots Z_p$ represent p stiffness requirements and 1/ $_1$ ...1/ $_p$ are the corresponding Lagrangian multipliers. Now the minimum condition is given by

$$\frac{\partial \phi}{\partial A_i} = 0 = \rho_i l_i + \frac{1}{\lambda_1} \frac{\partial}{\partial A_i} Z_1 + \cdots + \frac{1}{\lambda_p} \frac{\partial Z_p}{\partial A_i} \tag{15}$$

From Equation (15) the optimality condition can be written as

$$\frac{e_i^{(1)}}{\lambda_1} + \frac{e_i^{(2)}}{\lambda_2} + \cdots + \frac{e_i^{(p)}}{\lambda_p} = 1 \tag{16}$$

where $e^{(1)}_i \ldots e^{(P)}i$ are the ratio of the average strain energy to the mass density of the ith element in the corresponding design condition.

In Equation (16) $_1 \ldots$ $_p$ may be considered as weighting factors. These weighting factors are in general not a unique set and can be determined only approximately in an iterative procedure.

Case 2: Minimum Weight Structure for a Specified Displacement

As before the vectors R and r are the generalized force and displacement vectors. Now the stiffness requirement is specified by a single displacement component r_j, of the vector $\underline{r}$. A unit load vector with a nonzero component in the direction of the constrained displacement only is denoted by $\underline{F}$, i.e. the component $F_J = 1$ while all other elements of $\underline{F}$ are zero.

The virtual work of the unit load vector and the actual displacement vector are given by

$$Z = \underline{F}^t \underline{r} \tag{17}$$

Now constraining Z is equivalent to imposing a constraint on a specified displacement component.

Equations (2), (3), (17) and the minimum condition (10) once again give the Lagrangian multiplier in the form

$$\lambda = e_i \tag{18}$$

Now e_i is the ratio of the average virtual strain energy density to the mass density of the ith element and it is given by

$$e_i = \frac{\underset{\sim}{f}^t \underset{\sim}{K}_i \underset{\sim}{r}}{\rho_i v_i} \tag{19}$$

where $\underline{f}$ is the displacement vector corresponding to the unit load vector $\underline{F}$.

Now the statement of the optimality criterion for a single diplacement constraint is as follows:

"The optimum structure for a specified displacement is the one in which the ratio of average virtual strain energy density to the mass density is the same for all its elements".

The iterative algorithms derived in [4] and [7] lead to the designs satisfying this optimality criterion.

In the case of multiple constraints the optimality condition may be written as

$$\frac{e_i^{(1)}}{\lambda_1} + \frac{e_i^{(2)}}{\lambda_2} + \cdots\cdots + \frac{e_i^{(p)}}{\lambda_p} = 1 \tag{20}$$

where $e^{(1)}_i \ldots e^{(p)}_i$ are the ratios of average virtual strain energy density to the mass density of the ith element in the corresponding design condition.

It is interesting to note that while the strain energy densities are always positive, the virtual strain energy densities can be positive or negative quantities. This condition may sometimes provide information as to what members are to be eliminated for achieving the optimality condition.

Case 3: Minimum Weight Structure for Dynamic Stiffness

The dynamic stiffness requirement in the jth mode is expressed as

$$Z = \omega_j^2 = \frac{\underset{\sim}{X}_j^t \underset{\sim}{K} \underset{\sim}{X}_j}{\underset{\sim}{X}_j^t \underset{\sim}{M} \underset{\sim}{X}_j} \tag{21}$$

where $\underline{K}$ and $\underline{M}$ are the stiffness and mass matrices of the structure and ω_j is the jth natural frequency.

Substitution of Equation (21) in Equation(10)gives the expression for the Lagrangian multiplier in the following form:

$$\lambda = \frac{e_i}{\underset{\sim}{X}_j^t \underset{\sim}{M} \underset{\sim}{X}_j} \tag{22}$$

where e_i is the ratio of the difference in strain energy and kinetic energy densities to the mass density of the ith element and is given by

$$e_i = \frac{[\underset{\sim}{X}_j^t \underset{\sim}{K}_i \underset{\sim}{X}_j - \omega_j^2 \underset{\sim}{X}_j^t \underset{\sim}{M}_i \underset{\sim}{X}_j]}{\rho_i v_i} \tag{23}$$

where K and M. are the stiffness and mass matrices of the ith element in the structure coordinate system. In Equation (22) the denominator is the same for all the elements and the optimality criterion for the dynamic stiffness requirement may be stated as follows:

"A structure with a given natural frequency will be of minimum weight design when the ratio of the difference in strain energy density and kinetic energy density to its mass density is the same for all its elements while vibrating in its natural mode".

The above optimality criterion is identical to the one stated in [8] when the structure is built from the same material.

Case 4: Minimum Weight Structure With Stiffness Requirements for Static Stability

The stiffness requirement for static stability in the jth mode is expressed as

$$Z = \omega_j = \frac{\underset{\sim}{X}_j^t \underset{\sim}{K} \underset{\sim}{X}_j}{\underset{\sim}{X}_j^t \underset{\sim}{K}_G \underset{\sim}{X}_j} \tag{24}$$

where K and K_G are the stiffness and geometric stiffness matrices of the total structure and X_j is the mode of buckling. ω_j represents the jth critical buckling load. Equations (24), (7) and the minimum condition (10) give the expression for the Lagrangian multiplier in the following form:

$$\lambda = \frac{e_i}{\underset{\sim}{X}_j^t \underset{\sim}{K}_G \underset{\sim}{X}_j} \tag{25}$$

where e_i is the ratio of the strain energy density of the ith element in the buckled mode to its mass density and is given by

$$e_i = \frac{\underset{\sim}{X}_j^t \underset{\sim}{K}_i \underset{\sim}{X}_j}{\rho_i v_i} \tag{26}$$

In Equation (25) the denominator is the same for all elements and the optimality criterion for static stability requirements may be stated as follows:

"The optimum structure with stiffness requirements for static stability is the one in which the ratio of the average strain energy density to its mass density is the same for all its elements in the buckled mode".

The optimality criteria for multiple stiffness requirements in cases 3 and 4 may be expressed as before in the following form:

$$\frac{e_i^{(1)}}{\lambda_1} + \frac{e_i^{(2)}}{\lambda_2} + \cdots + \frac{e_i^{(p)}}{\lambda_p} = 1 \tag{27}$$

The quantities $e^{(l)}_i$ .... $e^{(p)}_i$ are given in Equations (23) and (26) for cases 3 and 4 respectively.

In conclusion it should be pointed out that the criteria derived in this section are valid for both statically determinate and indeterminate structures. However, in the case of indeterminate structures it is possible to obtain more than one design satisfying the optimality criterion.

III. Recursion Relations Based on Optimality Criteria

Except in the case of statically determinate structures the optimality criteria derived in the previous section can be attained by iteration only. The necessary recursion relations for iteration are derived in this section for all four cases.

Case 1. <u>Recursion Relation for Generalized Stiffness</u>

From Equations (12) and (13) the expression for the Lagrangian multiplier may be written as

$$\lambda = \frac{u_i}{A_i \; l_i \; \rho_i} \tag{28}$$

where u_i is the total strain energy in the ith element. The design variable vector $\underline{A}$ is written as

$$\underset{\sim}{A} = \Lambda \, \underset{\sim}{\alpha} \tag{29}$$

where Λ is the normalizing or scaling parameter and $\underline{\alpha}$ is the relative design variable vector. From Equations (28) and (29) the expression for Λ may be written as

$$\Lambda^2 = \frac{1}{\lambda} \left[u_i' \; / \; \tau_i' \right] \tag{30}$$

where u_i' and τ_i' are given by

$$u_i' = \frac{1}{2} \underset{\sim}{r}^t \, \underset{\sim}{K}_i \, \underset{\sim}{r} \, \Lambda \tag{31}$$

$$\tau_i = \Lambda \, \tau_i' = \Lambda \, \alpha_i \, l_i \, \rho_i \tag{32}$$

Multiplying both sides of Equation (30) by α_i^2 and taking the square root, the expression for the design variable may be written as

$$\alpha_i \Lambda = C \alpha_i [u_i' / \tau_i']^{\frac{1}{2}} \quad (33)$$

Since α_i **exists** on both sides of the equation the recursion relation may be written as

$$(\alpha_i \Lambda)_{\nu+1} = C (\alpha_i)_\nu [u_i' / \tau_i']_\nu^{\frac{1}{2}} \quad (34)$$

where ν represents the cycle of iteration.

The generalized stiffness Z of the structure may be written as

$$Z = \sum_{i=1}^{m} A_i l_i e_i \rho_i \quad (35)$$

From the optimality condition (Equation (12)) Equation (35) may be written as

$$Z = \lambda W \quad (36)$$

where **W** is the weight of the structure. Substitution of Equation (36) in (34) gives the final recursion relation in the form

$$(\alpha_i \Lambda)_{\nu+1} = (\alpha_i)_\nu [W/Z]_\nu^{\frac{1}{2}} [u_i' / \tau_i']_\nu^{\frac{1}{2}} \quad (37)$$

A step by step procedure for utilizing the recursion relation (37) is outlined in Section 5.

In case of multiple stiffness requirements, such as those arising from independent loading conditions and other stiffness requirements, any procedure for determining the weighting parameters has to be necessarily approximate. The weighting parameters are not a unique set in the sense that more than one set of parameters can satisfy the optimality condition given by Equation (16). From Equation (27) an obvious candidate for the recursion relation in case of multiple stiffness requirements is

$$(\alpha_i \Lambda)_{\nu+1} = (\alpha_i)_\nu \left[c_1 (u_i^{(1)'}/\tau_i') + \cdots + c_p (u_i^{(p)'}/\tau_i')\right]_\nu^{\frac{1}{2}} \tag{38}$$

where the weighting parameters $c_1 \ldots c_p$ are given by

$$c_j = \frac{W}{Z_j} \tag{39}$$

where **W** is the total weight of the structure.

<u>Case 2: Recursion Relation for Prescribed Displacements</u>

As in the previous case the recursion relation for prescribed displacements can be written as

$$(\alpha_i \Lambda)_{\nu+1} = c\, (\alpha_i)_\nu \left[u_i' / \tau_i'\right]_\nu^{\frac{1}{2}} \tag{40}$$

where u_i' in this case is given by

$$u_i' = \tfrac{1}{2}\, \underset{\sim}{f}^t\, \underset{\sim}{K}_i\, \underset{\sim}{r}\, \Lambda \tag{41}$$

where $\underline{f}$ is the displacement vector resulting from the unit load vector corresponding to the active displacement direction. In using Equation (40) it should be remembered that the active displacement direction may change from cycle to cycle and consequently the unit load vector may not be the same in all cycles even though there is only one active displacement at any given time. When u_i' is negative the member size can be placed in the passive group and the minimum size can be assigned to that member at least in that cycle. This situation may change in the next cycle in which case the member can be moved into the active group again.

The weighting constant C is given by

$$C = \frac{W}{Z} = \frac{W}{\underset{\sim}{F}^t \, \underset{\sim}{r}} \tag{42}$$

In case of multiple active constraints the proposed iterative algorithm is as follows:

$$(\alpha_i \, \Lambda)_{\nu+1} = (\alpha_i)_\nu \, [C_1 (u_i^{(1)'} / \tau_i') + \dots + C_p (u_i^{(p)'} / \tau_i')]_\nu^{\frac{1}{2}} \tag{43}$$

where the weighting parameters C_j are given by

$$C_j = W/Z_j \tag{44}$$

<u>Case 3: Recursion Relation for Dynamic Stiffness</u>

The iterative algorithm for dynamic stiffness requirements is as follows:

$$(\alpha_i \, \Lambda)_{\nu+1} = C \, (\alpha_i)_\nu \, [u_i' \, / \, \tau_i']_\nu^{\frac{1}{2}} \tag{41}$$

where the proportionality constant C is given by

$$C = \frac{W}{Z-1} \; \frac{1}{\underset{\sim}{X}_j^t \underset{\sim}{M} \underset{\sim}{X}_j} \tag{45}$$

u_i' and τ_i' are given by

$$u_i' = \tfrac{1}{2}\left[\underset{\sim}{X}_j^t \underset{\sim}{K}_i \underset{\sim}{X}_j - \underset{\sim}{X}_j^t \underset{\sim}{M}_i \underset{\sim}{X}_j\right]\Lambda \tag{46}$$

$$\tau_i = \Lambda\, \tau_i' = \Lambda\, \alpha_i\, l_i\, \rho_i \tag{47}$$

<u>Case 4: Recursion Relation for Stiffness for Static Stability</u>

From Equations (21) and (22) the recursion relation for static stability stiffness is given by

$$(\alpha_i\, \Lambda)_{\nu+1} = C\, (\alpha_i)_\nu\, [u_i' / \tau_i']_\nu^{\frac{1}{2}} \tag{48}$$

where the constant of proportionality C is given by

$$C = \frac{W}{Z} \; \frac{1}{\underset{\sim}{X}_j^t \underset{\sim}{K}_G \underset{\sim}{X}_j} \tag{49}$$

The expressions for u_i' and τ_i' are similar to those given by Equations (31) and (32).

IV-Designs Based on Optimality Criteria

The optimality criteria derived in Section 2 cover most design situations with stiffness requirements. In case of determinate structures these criteria represent both necessary and sufficient conditions for a global minimum for the specified stiffness requirements. In indeterminate structures, however, there may be a number of designs with the same weight or different weights but satisfy both the stiffness requirements and the optimality conditions. Each of these designs represents a relative minimum. This nonuniqueness and the necessity to iterate in the case of indeterminate structures makes every design method an approximate procedure. For effective use of the optimality criteria approaches the designer must be aware of certain characteristics of these designs. The remainder of this section points out some of these characteristics with the aid of a two variable design space.

Figure 1 represents a two variable design space. Suppose Point A in this space represents a design that satisfies the optimality criterion, then it can be shown under certain conditions that every design on line OA and its extension satisfy the optimality criterion. The line OA may be called the optimality criterion line. Two possibilities exist:

1) If the line OA is unique (i.e. no other lines satisfy the optimality criterion), then the lowest feasible design on line OA is the global minimum.

2) If more than one optimality criterion line exists, then corresponding to each line there will be a lowest feasible design that represents a relative minimum.

For determinate structures it is always possible to obtain a design satisfying the optimality criterion. For indeterminate structures, however, two possibilities exist:

1. No design with all members present can satisfy the optimality criterion, i.e. the indeterminate structure degenerates to one of the determinate cases when the optimality criterion is enforced.

2. An indeterminate structure satisfying the optimality criterion

can be found. But there will be at least one or more determinate structures which will also satisfy the optimality criterion.

The optimality criteria derived in Section 2 contain only certain stiffness requirements and no other design constraints are imposed. In practice, however, design requirements include constraints on stresses and sizes of the elements. In general these requirements are inequality constraints and they are too numerous to include in any Lagrangian multiplier type of formulation. In such situations a distinction between an optimum design and a design that satisfies the optimality criterion is essential for effective use of the iterative algorithms derived in Section 3. For instance in Figure 2 the point C_n represents the design that satisfies the optimality criterion. Assuming that the iteration started from C_1, it reached C_n in n steps (represented by the radial lines). In an ideal case, when none other than the stiffness requirements are present, the optimum design and the design that satisfies the optimality criterion are both represented by the same line OC . In a more practical case this path is interrupted at point I by some additional constraints, and the constraint surface is shifted to the dotted line and the design ends at C_n' where the optimality criterion is satisfied. Here I represents the optimum design while C_n' represents the design satisfying the optimality criterion. Since the object is to find the optimum and not satisfaction of the optimality criterion the design should be terminated at I and should not be forced to C_n'. The usefulness of the optimality criterion approach to optimum design is due to the fact that, in general, the optimum point lies on the path to the point that satisfies the optimality criterion.

V. Design Procedure and the Examples

The design examples presented in this section are grouped into four categories. The first group is designed for static loads with constraints on stresses, displacements and minimum sizes. The second group of examples are designed for dynamic loads with periodic time dependence. The third group consists of layered composite structures subjected to static loads. The fourth group are examples with static stability requirements.

The step by step procedure for the use of the iterative algorithms derived in Section 3 for the static case is as follows:

1. The structure is analyzed with the initial relative design vector consisting of equal sizes for all members.

2. The constraint surface is located by scaling the design to satisfy the stress and displacement constraints.

3. The weight of the feasible design is determined.

4. The active displacement constraints are identified and the response of the structure for the unit load vectors in the direction of the active constraints is determined.

5. The structure is resized by using the iterative algorithms derived in Section 3.

6. Steps 1 thru 5 are repeated so long as the design improves.

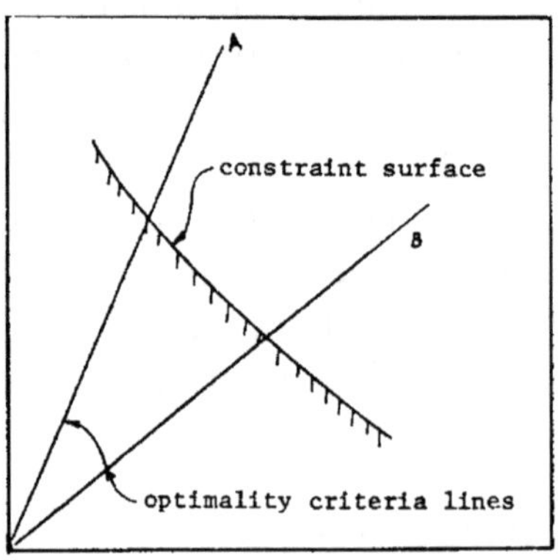

Fig. 1. Two variable design space

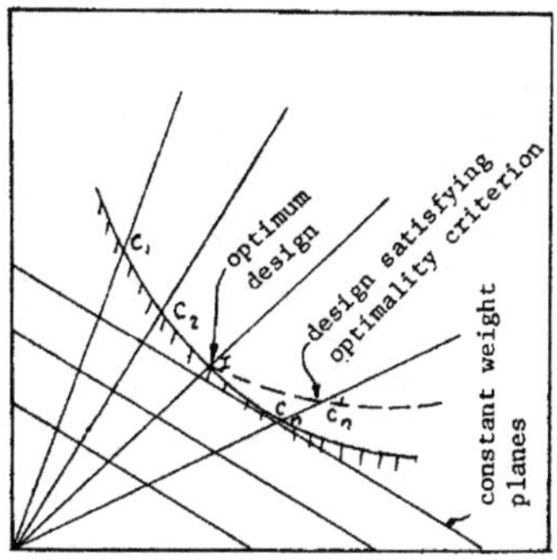

Figure 2. Optimum and Optimality Criteria Designs

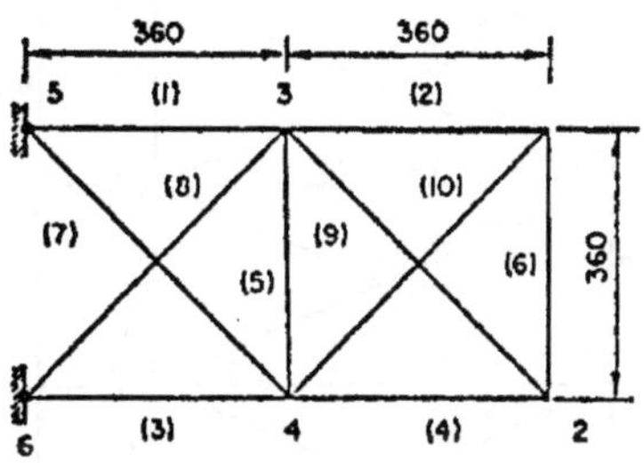

Fig. 3 Cantilever truss

Table 1. Design Data for Examples 1-9

Example	1,2,3*	4,5,6	7	8	9
El. Mod. E(10^6 Lbs/in^2)	10.000	29.000	29.000	29.000	10.000
Min. Siz. (in^2)	0.1				0.1
Rad. of Gyrn. (in)		2.646	3.426	8.731	
Ms. Dnsty. (10^{-2} slugs/in^3)	0.1**	0.884		0.888	0.313
Strs. Lmts. (10^3 Lbs/in^3)	25.000		0.884	29.000	25.000

*For Example 3 stress limits for individual elements are given in the text.

**lbs/in^3.

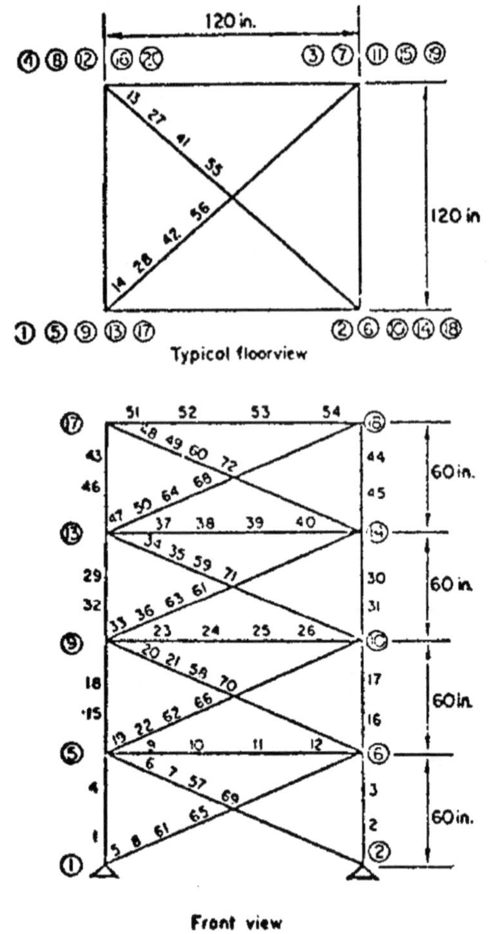

Fig. 4 Four level tower

Design Examples for Static Case

Example 1: Cantilever Truss: Single Loading: Stress and Displacement Constraints

The cantilever truss shown in Figure 3 is designed for two different loading cases. In both cases there are stress and displacement constraints. Designs for this truss were presented earlier in [3] and [7]. The present designs are slight improvements over the ones given in the past.

Additional Design Information:

Displacement Limits ±2.0 inches in y-direction on all nodes

Loading Information

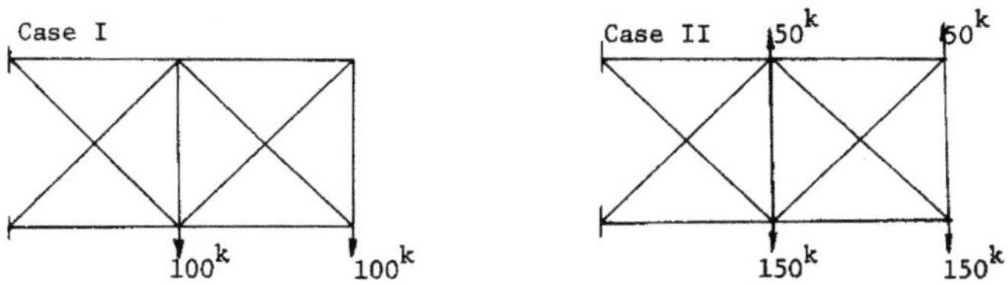

Cantilever Truss Design for First Case:

In 1970 a design weighing 5084.9 lbs was reported in [3] for this structure. Reference [7] contains a design weighing 5112 lbs. The present design of 5069.19 lbs is the lowest weight design found so far for this structure. The procedure for finding this design is the same as the one given in [3] except for adjustment of the step size in the iteration. The details of the three designs are given in Table 2.

Cantilever Truss Design for Second Case:

For this problem [3] contains a design weighing 4895.6 lbs. Subsequently the design given in table 3 was found by adjustment of the step size in the iteration. The present design is about 183 lbs (about 4%) lighter than the one reported in [3]. The details of the two designs are given in Table 3.

Example 2: Four Level Tower: Stress and Displacement Limits

The design of this structure was first reported in [2]. Subsequently the same structure was used as a test problem in [3], [5], [7] and [12]. The structure is subjected to five loading conditions. Four of these are devised to maintain symmetry of the structure. Details of the geometry are given in Figure 4 and the loading details are given in Table 4. Table 5 contains the details of the designs given in [2], [3], [5], [7] and [12]. It is interesting to note that the designs presented in [3], [5] and [12] are not significantly different in merit even though they differ in individual sizes of the elements.

Cantilever Truss

Design for First Case — Table 2; Design for Second Case — Table 3

Memb.	Sizes of the Members (Table 2)			Sizes of the Members (Table 3)	
	Ref. 3	Ref. 7	New	Ref. 3	New
1	30.416	31.35	30.882	25.190	23.737
2	0.128	0.10	.100	0.363	0.100
3	23.408	20.03	23.509	25.419	24.953
4	14.904	15.60	14.930	14.327	13.961
5	0.101	0.14	.100	0.417	0.100
6	0.101	0.24	.441	3.144	3.007
7	8.696	8.35	7.483	12.083	12.712
8	21.084	22.21	21.266	14.612	13.438
9	21.077	22.06	21.114	20.261	19.744
10	0.186	0.10	.100	0.513	0.100
Weight	5084.9	5112.0	5062.3	4895.6	4712.5

Loading Conditions (Example 2)

Table 4

Load cond.	Node	Direction of Load x	y	z
1	17	5000	5000	-5000
2	18	-5000	5000	-5000
3	19	-5000	-5000	-5000
4	20	5000	-5000	-5000
	17	0	0	-5000
5	18	0	0	-5000
	19	0	0	-5000
	20	0	0	-5000

Example 3: Three Bar Truss: Stress Limits

The design of this truss was first reported in *[13]* and was subsequently used by several investigators as a test problem. The results given in [3], *[12]* and *[13]* are reproduced along with the new design in Table 6 for comparison purposes. Interestingly the final designs are not significantly different in merit even though the difference in individual sizes are somewhat more pronounced.

Four Level Tower

Table 5

Members	Sizes of the Members					
	Ref.2	Ref.3	Ref.5	Ref.7	Ref.12	New
1,2,3,4	1.471	1.818	1.635	1.464	1.897	1.818
5,6,7,8	0.548	0.523	0.525	0.521	0.516	0.523
9,10,11,12,13,14	0.090	0.100	0.105	0.100	0.100	0.100
15,16,17,18	1.083	1.246	1.254	1.024	1.280	1.248
19,20,21,22	0.559	0.524	0.524	0.542	0.515	0.525
23,24,25,26,27,28	0.090	0.100	0.112*	0.100	0.100	0.100
29,30,31,32	0.933	0.611	0.684	0.552	0.507	0.610
33,34,35,36	0.589	0.532	0.527	0.608	0.520	0.532
37,38,39,40	0.137	0.100	0.110	0.100	0.100	0.100
41,42	0.112	0.100	0.178	0.100	0.100	0.100
43,44,45,46	0.858	0.161	0.154	0.149	0.157	0.160
47,48,49,50	0.674	0.557	0.542	0.773	0.536	0.559
51,52,53,54	0.518	0.377	0.469	0.453	0.410	0.374
55,56	0.503	0.506	0.491	0.342	0.564	0.500
57,61,65,69	0.548	0.523	0.525	0.521	0.516	0.523
58,62,66,70	0.559	0.524	0.524	0.542	0.515	0.525
59,63,67,71	0.589	0.532	0.527	0.608	0.520	0.532
60,64,68,72	0.674	0.557	0.542	0.773	0.536	0.559
Weight	425.8	381.2	384.9	395.97	379.6	381.1

*Members 27 and 28 are 0.105 in^2

Iteration History

Cycle	1	2	3	4
Weight	656.8	416.2	382.5	381.1
Computer Time CDC 6600 5 Sec.				

Ref.	Areas of Members			Weight in lbs
	1	2	3	
13	7.099	1.849	2.897	16.0
3	7.116	1.790	2.925	16.0
12	7.014	2.178	2.737	16.0

Member	1	2	3
Stress Limits	5000	20,000	5000

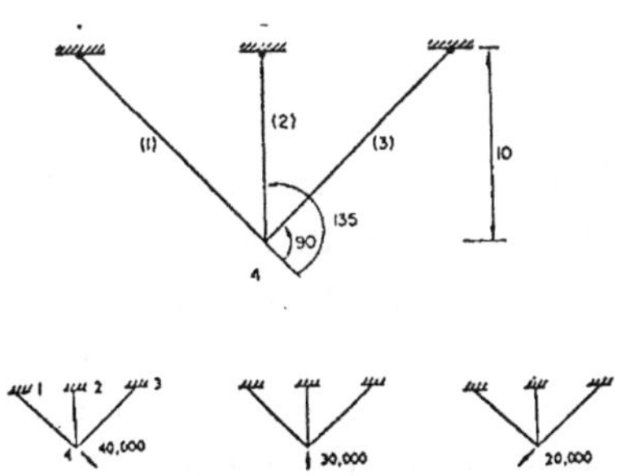

Fig. 5 Three bar truss

<u>Design Examples for Dynamic Case</u>

The optimality criterion derived in Section 2 is for structures vibrating in their natural modes. However, the procedure is extended to designing in the dynamic mode [8]. The dynamic mode may be a single natural mode of the structure or a linear combination of a set of natural modes depending on the spatial distribution and dynamic characteristics of the forcing function. The dynamic mode for an arbitrary forcing function $\underline{R}$ (t) is written as

$$\underset{\sim}{r} = \underset{\sim}{\Psi}\,\underset{\sim}{\Omega}\,\underset{\sim}{D}\,\underset{\sim}{\Psi}^{t}\,\underset{\sim}{R}_{o} \tag{50}$$

where ψ is the modal matrix consisting of critical modes. The procedure for filtering the critical modes is discussed in [8]. It is based on the study of the variation of the virtual work of the peak forces with the number of modes included in the dynamic response. $\underline{R}_o$ is the vector of peak forces. $\underline{\Omega}$ and $\underline{D}$ are diagonal matrices with elements given by

$$\Omega_i = \frac{1}{\omega_i^2 M_i} \tag{51}$$

$$D_i(t) = \int_0^t \omega_i \sin[\omega_i(t-\tau)]\, F(\tau)\, d\tau \tag{52}$$

where t represents the time at which the response is designed and T is the intermediate time variable. For additional details the reader is referred to [8].

Examples 4,5 and 6 are for the purpose of studying the variation of the natural frequencies and material distribution when the structures are designed in their natural modes. Designing in one of the natural modes is tantamount to designing in the actual dynamic mode if the given forcing function activates only that natural mode. The scaling of the design to the constraint surface is the only additional step required to obtain the actual design. For instance if a loading condition activates only the fundamental mode, then the relative distribution of the material obtained by designing in the fundamental mode would be the same as that obtained by designing in the actual dynamic mode. In Examples 7 and 8 the structures are designed for periodic dynamic forces. The rectangular frame in example 7 is designed for three different loading cases. These loading cases are devised to show how the spatial distribution of the forcing function activates different modes even though the dynamic characteristics are the same in all cases. The 8th example is a wing structure idealized by bar elements. The structure is designed for a periodic forcing function. It represents a medium size problem in terms of degrees of freedom and number of elements.

In the examples involving beam elements the principal moment of inertia is assumed to be the design variable. It is also assumed that the radius of gyration of the beam section is the same for all the elements. A sandwich beam with constant depth and variable face sheet thicknesses would conform to this assumption. The elastic constants, mass properties and other design data are given in Table 1.

Example 4: Cantilever Beam - Design in the First Mode.

The cantilever beam shown in Figure 6 is designed in its fundamental mode. The design started with the assumption of equal sizes for all the elements. The elements are resized by the iterative algorithm based on the optimality criterion (Equation (44)). The first few natural frequencies, the relative weight, and the ratio of the fundamental frequency to the relative weight are given in Table 7 for each iteration. This table also contains relative sizes of the elements in the final design. The distribution of the moment of inertia along the length of the beam is shown in Figure 6.

Example 5: Simply Supported Beam: Design in the First Mode

The simply supported beam shown in Figure 7 is designed in its fundamental mode. The first three natural frequencies, the relative weight of the beam, and the ratio of the fundamental frequency to the relative weight are given in Table 8 for each iteration. It is interesting to note that the values of the natural frequencies are

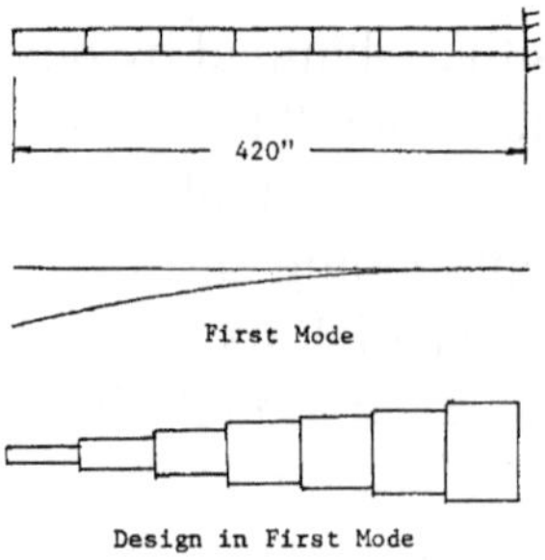

Figure 6 Cantilever Beam

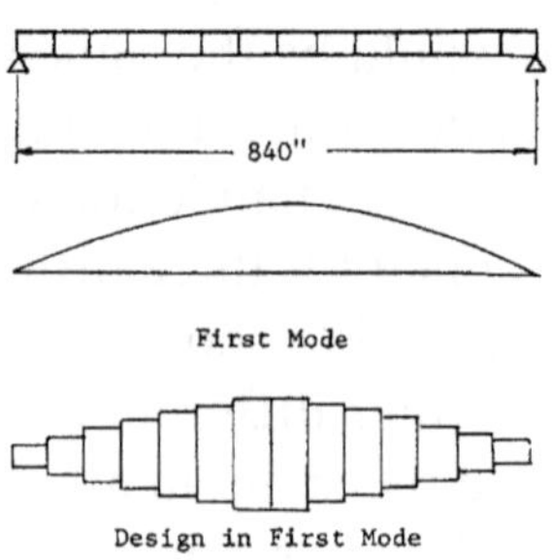

Figure 7 Simply Supported Beam

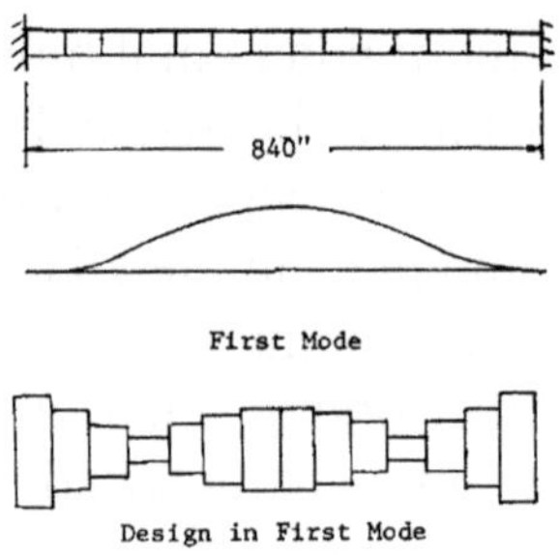

Figure 8 Fixed-Fixed Beam

TABLE 7 CANTILEVER BEAM

Cycle	λ_1	λ_2	λ_3	Rel.Wt.	λ_1/Wt.
1	.316	12.312	97.680	16.966	.019
2	2.353	31.270	150.098	7.413	.317
3	5.043	46.914	179.102	5.885	.857
4	7.398	58.164	200.195	5.280	1.401
5	9.086	65.195	210.742	4.991	1.820
6	10.264	70.117	221.289	4.840	2.121
7	10.967	72.578	224.805	4.754	2.307
8	11.494	75.391	228.320	4.704	2.443
9	11.846	76.797	231.836	4.673	2.535
10	12.022	76.797	231.836	4.653	2.584

SIZES OF THE ELEMENTS IN THE FINAL DESIGN

Elem	1	2	3	4	5	6	7
Size	.0002	.003	.018	.082	.250	.566	1.00

$\lambda_i = \omega_i^2 \, 10^6 / E^*$

λ_1	λ_2	λ_3	Rel.Wt.	λ_1/Wt.
.155	2.494	12.549	33.932	.005
.145	2.248	11.846	21.921	.007
.143	2.213	11.846	21.402	.007
.143	2.213	11.846	21.365	.007
.143	2.213	11.846	21.362	.007

SIZES OF THE ELEMENTS IN THE FINAL DESIGN

1	2	3	4	5	6	7
.119	.313	.508	.687	.836	.944	1.00

TABLE 8 SIMPLY SUPPORTED BEAM

TABLE 9 FIXED - FIXED BEAM - DESIGN IN FIRST MODE

Cycle	λ_1	λ_2	λ_3	λ_4	Rel.Wt.	λ_1/Wt.
1	.796	6.027	23.183	63.789	33.933	.023
2	.993	8.242	33.379	76.797	18.034	.055
3	1.009	8.102	31.973	72.578	18.231	.055
4	1.044	8.102	31.621	70.117	17.879	.058
5	1.062	8.242	31.621	69.414	17.691	.060
6	1.062	8.242	31.621	69.414	17.597	.060
7	1.062	8.242	31.621	69.414	17.553	.061
8	1.062	8.242	31.270	68.711	17.530	.061
9	1.062	8.242	31.270	68.711	17.518	.061
10	1.062	8.242	31.270	68.711	17.512	.061

Elem	1	2	3	4	5	6	7
Size	1.00	.657	.330	.089	.307	.545	.684

TABLE 10 RECTANGULAR FRAME (DESIGNS IN FOUR MODES)

		λ_1	λ_2	λ_3	λ_4	Rel.Wt.
Init Des		.413	1.343	10.440	24.590	34.700
Final Design	Mode					
	1	1.062	1.642	9.508	31.270	11.227
	2	.175	.979	9.367	23.184	16.198
	3	.164	.852	8.242	26.699	15.026
	4	.589	1.896	11.494	22.832	14.866

$\lambda_i = \omega_i^2\ 10^6/E^*$

SIZES OF THE ELEMENTS IN THE FINAL DESIGN

	Elem	1	2	3	4	5	6	7	8	9
Mode	1	1.000	.733	.475	.234	.066	.236	.292	.198	.079
	2	.209	.062	.154	.316	.473	.620	.366	.518	1.000
	3	.183	.060	.125	.231	.288	.293	.433	1.000	.574
	4	1.000	.267	.621	.947	.722	.219	.334	.180	.169

SYMMETRICAL ABOUT CENTER LINE

reduced slightly but the reduction in weight is much more pronounced. The relative sizes of the elements in the final design are given in Table 8. The graphic representation of the material distribution is shown in Figure 7.

Example 6: Fixed - Fixed Beam: Design in the First Mode

The fixed-fixed beam shown in Figure 8 is designed in its fundamental mode. The results of this design are given in Table 9 and the graphic representation of the material distribution is shown in Figure 8.

Example 7: Rectangular Frame: Design in the First Four Modes

The rectangular frame shown in Figure 9 is designed independently in the first four modes. The natural frequencies and the relative sizes are given in Table 10. The natural modes and the material distribution of the final designs are shown qualitatively in Figure 9.

Example 8; Rectangular Frame: Design for Dynamic Loads

The rectangular frame shown in Figure 9 is designed for three different cases of dynamic loads. In all three cases the forcing functions are periodic and have the same circular frequency. The spatial distribution of the forcing functions is shown for the three cases in Figure 10.

The dynamic load factor, $D_r(t)$, corresponding to the periodic forcing function after integration (Equation 52) is given by

$$D_r(t) = \frac{1}{[1 - (\frac{p}{\omega_r})^2]} (\sin pt - \frac{p}{\omega_r} \sin \omega_r t) \tag{53}$$

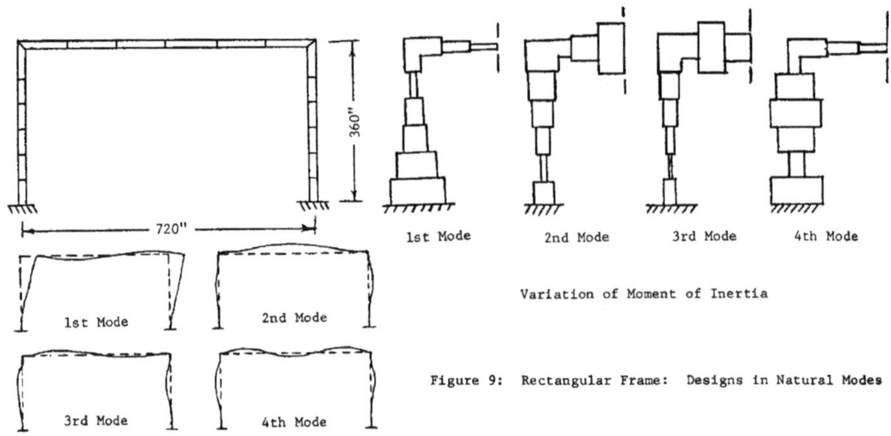

Figure 9: Rectangular Frame: Designs in Natural Modes

The peak values of $D_r(t)$ are used in the dynamic load factor matrix in evaluating response (Equation (50)) even though they do not occur at the same time. This approximation, in general, produces conservative designs. A more accurate procedure would be to evaluate the dynamic response at sufficiently small time intervals and to select the response at which the dynamic stresses are a maximum.

Case 1: Dynamic Lateral Force

In the first case the frame is subjected to a periodic lateral force. The design started with the assumption of equal sizes for all the elements. For this design a plot of the virtual work, V_W, against the number of modes in the response is given in Figure 11. This plot shows that the first mode is the most significant one, and the modes beyond the fourth have little contribution. The first four natural frequencies of the initial and the final designs for this loading case are given in Table 11. Table 11 also contains the sizes of the elements in the final design. The graphic representation of the material distribution is shown in Figure 12.

Case 2: Dynamic Vertical Force

The plot of the virtual work, V_w , for this Case (Figure 11) shows that only the second mode contributes significantly to the dynamic response. All other modes including the first have very little contribution. The final design and the first four natural frequencies are given in Table 11. The material distribution is shown in Figure 12.

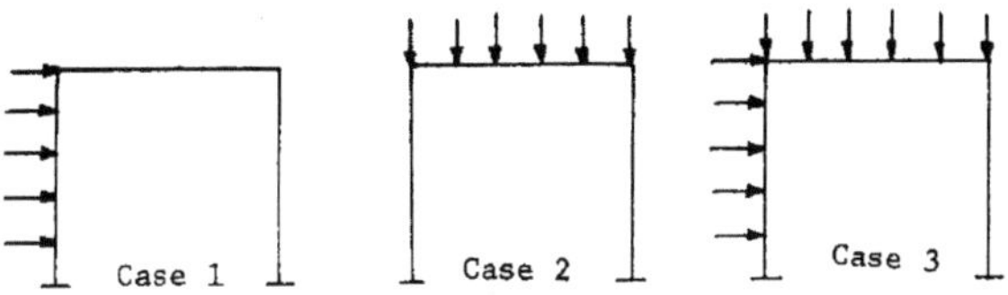

Peak Forces

$R = R_o \text{ Sin } pt$

Case 1	1000 Lbs at Nodes 2 to 7 (X-Direction)
Case 2	1000 Lbs at Nodes 7 to 13 (Y-Direction)
Case 3	Combination of Cases 1 and 2

*Note: p/ω_r <<1/10 in all cases

Figure 10. Dynamic Load Distribution

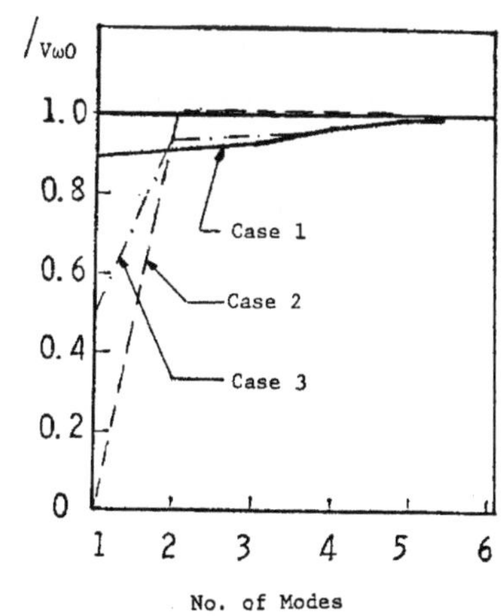

Figure 11. Virtual Work ~ Modes in Dynamic Response

Table 11 Rectangular Frame: Design for Dynamic Loads

Case 1: Periodic Lateral Forces $\lambda_i = \omega_i^2 \, 10^6/E^*$ Case 2: Periodic Vertical Forces

Design	λ_1	λ_2	λ_3	λ_4	Weight	λ_1	λ_2	λ_3	λ_4	Weight
Initial	2.67	8.66	67.30	155.37	1080	2.67	8.66	67.31	155.37	563
Final	6.52	12.73	71.17	214.26	522	1.57	7.54	63.79	151.86	506

Cycle No.	1	2	3	4
Weight	1080	808	528	522

Cycle No.	1	2	3	4
Weight .	563	534	517	506

Elem.	1	2	3	4	5	6	1	2	3	4	5	6
Area	6.61	3.75	1.56	0.36	0.99	1.18	0.77	0.35	0.48	0.95	1.48	2.02
Inertia	503.80	285.51	119.11	27.77	75.43	90.05	58.71	26.40	36.29	72.36	112.61	154.19
Elem.	7	8	9	10	11	12	7	8	9	10	11	12
Area	0.91	0.53	0.19	0.36	0.76	1.16	1.22	1.13	2.06	2.06	1.13	1.22
Inertia	69.64	40.16	14.23	27.64	57.64	88.47	92.68	85.76	157.17	157.17	85.76	92.68
Elem.	13	14	15	16	17	18	13	14	15	16	17	18
Area	0.99	0.30	0.63	1.39	2.15	2.92	2.02	1.48	0.95	0.48	0.35	0.77
Inertia	75.77	22.54	48.20	105.64	163.89	222.61	154.19	112.61	72.36	36.29	26.40	58.71

*E Given in Table 1

Case 3: Periodic Combined Forces

Design	λ_1	λ_2	λ_3	λ_4	Weight
Initial	2.67	8.66	67.31	155.37	940
Final	2.53	8.10	63.09	200.20	718

Cycle No.	1	2
Weight	940	718

Elem.	1	2	3	4	5	6
Area	3.41	1.73	0.65	0.26	0.26	0.82
Inertia	260.18	131.87	49.34	19.70	19.70	62.16
Elem.	7	8	9	10	11	12
Area	0.70	1.79	2.35	1.82	0.69	2.98
Inertia	53.40	136.49	178.89	138.87	52.90	227.35
Elem.	13	14	15	16	17	18
Area	3.83	2.30	0.90	0.98	2.41	3.91
Inertia	292.10	175.65	68.82	74.49	183.49	297.90

Table 12 Wing Structure: Design for Dynamic Loads

Design	λ_1	λ_2	λ_3	Weight
Initial	23.887	146.582	448.047	9220.312
Final	880.469	1043.95	3337.89	2115.030

Cycle No.	1	2	3	4
Weight	9220.312	3278.977	2389.517	2115.030

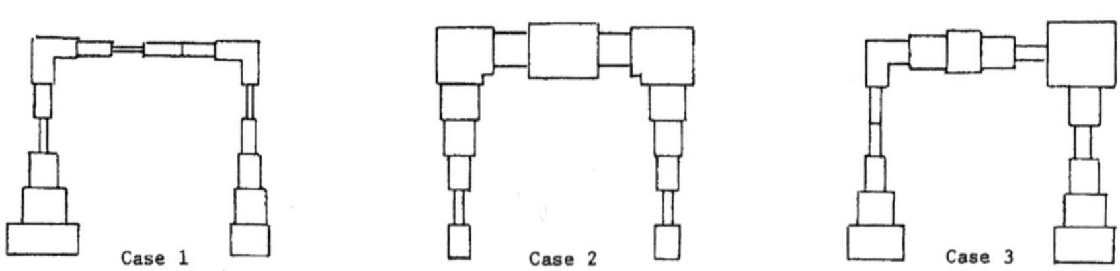

FIGURE 12. Material Distribution of Dynamically Loaded Frame

Rib Dimensions

Rib AT	AB	CD	EF	GH
X = 0"	5.25"	7.00"	5.25"	3.50
X = 600"	24.00"	32.00"	24.00"	16.00"

Loads Per Node (Pounds)

1	2	3
500	1000	1500

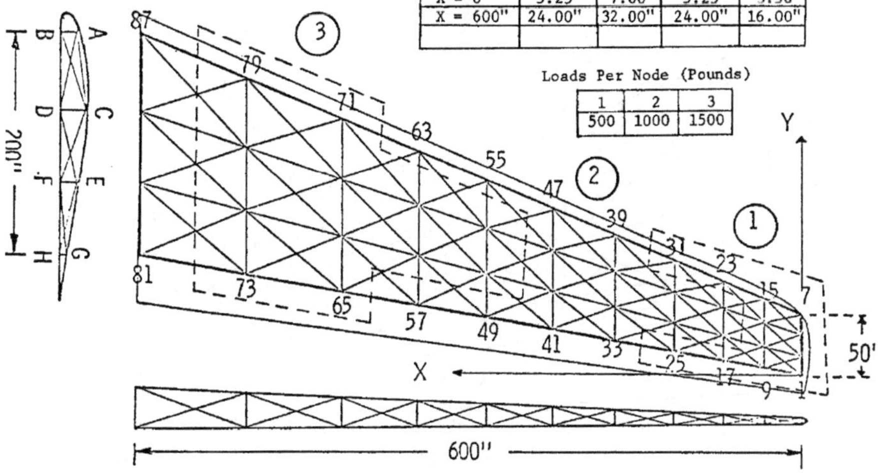

Figure 13 Wing Structure

Rib Dimensions

Rib AT	AB	CD	EF	GH
X = 0"	5.25"	7.00"	5.25"	3.50
X = 600"	24.00"	32.00"	24.00"	16.00"

Loads Per Node (Pounds)

1	2	3
500	1000	1500

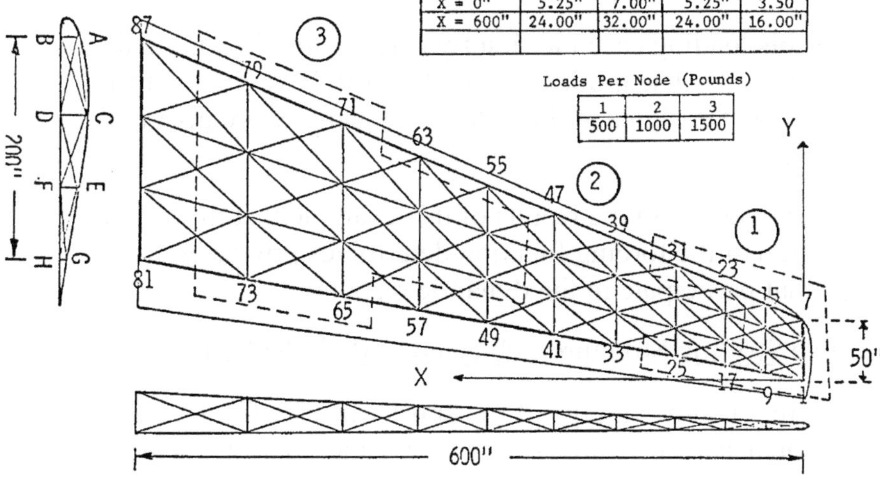

Figure 13 Wing Structure

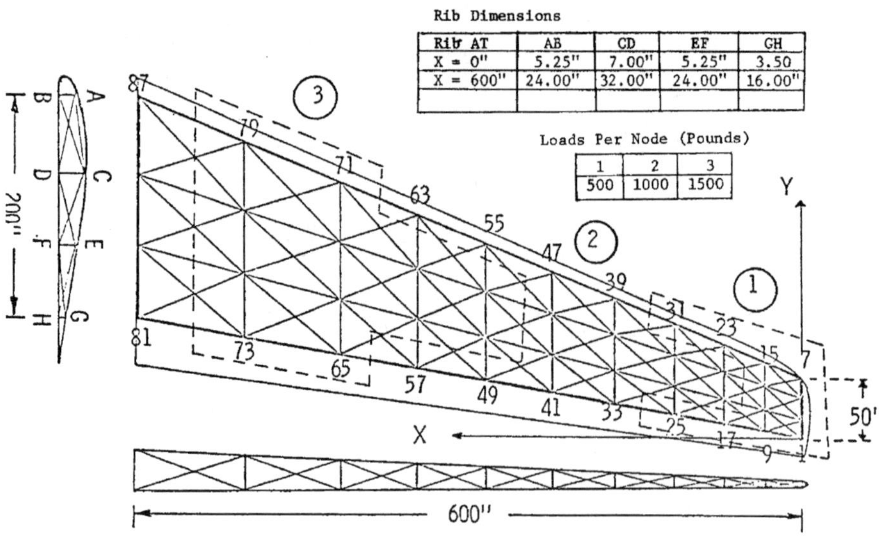

Figure 13 Wing Structure

Case 3: Combination of Lateral and Vertical Forces.

For this loading case the first two are the significant modes (See Figure 11). The modes beyond the fifth have little contribution. The final design and the natural frequencies are given in Table 11.

It is evident from the results of this example that the spatial distribution of the forcing function determines primarily the modes that participate in the dynamic response.

Example 9: Wing Structure Subjected to Periodic Forces

Figure 13 shows a typical transport wing idealized as a three dimensional bar structure. It has two hundred forty degrees of freedom and four hundred forty elements. The wing is subjected to a sinusoidal forcing function. The final design and the first three natural frequencies are given in Table 12.

Design Examples: Layered Composite Structures

The next two problems are concerned with minimum weight

design of structures made from fiber reinforced composite materials such as boron-epoxy or graphite-epoxy. The recurrence relation used in resizing elements at each iteration is derived in Section 3. For additional details the reader is referred to [9].

In structures made from isotropic materials, the material constants do not change during the design and the design variables are functions of geometric properties only. However, in composite structures, because of the change in the number of layers, and the fiber orientations in each layer, the material constants can change during the design. In an ideal optimization situation, there should be freedom to change the fiber orientation in each layer from point to point in the structure. However, manufacturing difficulties preclude such freedom and the general practice is to select a finite number of predetermined fiber orientations and to change the number of laminae in each of the directions to obtain an efficient structure. The directions 0^0, 90^0, $+45^0$ and -45^0 are generally chosen as basic fiber orientations.

A structural element in a general state of stress requires a strength criterion to determine its ability to withstand the stress resulting from the applied load. The strength criterion is one of thfactors that determines the magnitude of the scaling parameter A and it determines the location of the constraint surface in the design space. For isotropic elements, the Von Mises criterion is used, and for composite elements Hills criterion modified by Tsai is used. These criteria are selected for their simplicity, however a different criteria can be incorporated into the computer program with little effort. The computer program contains four types of finite elements (1) constant strain triangles (2) quadrilaterals constructed from four constant strain triangles (3) shear panels, and (4) bars. The triangles and quadrilaterals may be layered composite elements with fiber orientations in the 0^0, 90^0, $+45^0$ and -450 directions. The number of laminae in the $+45^0$ and -45^0 directions are kept equal by taking the average strain energies stored in both layers.

Two problems are presented here to illustrate the applicability of the approach. The problems are, a rectangular plate with a hole and a cantilever wing. The wing is subjected to two loading conditions. The initial or starting design for boron-epoxy members consist of equal percentages for each fiber orientation. The elastic constants, the intermediate steps and the complete details of the final designs are given in [9].

Example 10: Layered Composite Plate with a Hole

A 6"xl2" plate with a two inch diameter hole at the center is subjected to a tensile load of 15 kips/ inch parallel to the 12 inch side. For the finite element model only one quarter of the plate is considered because of the symmetry of the loading and the structure. The quarter plate idealization has 142 nodes and 138 plate elements (membrane quadrilaterals and triangles). The weight of the initial design (equal percentages) required to satisfy the strength criterion is 4.84 lbs and is reduced to 0.802 lbs. after eight iterations. The contours of the number of laminae in each fiber direction are given in Figure 14. The 0^0 fibers are in the direction of the load. The average CP time (CDC 6600) required for each iteration is 24.7 seconds.

Example 11: Cantilever Wing

The finite element model of a cantilever wing is shown in Figure 15. The structural model has 88 nodes and 170 members. The top and bottom skins are idealized by membrane quadrilaterals, and the spars and the ribs are idealized by shear panels and bars. The structure is constrained in all directions at the root section. The wing is optimized first with all aluminum elements. The design is started with the relative sizes of 1.0 for bar areas and 0.1 for shear and membrane element thicknesses. The wing is next optimized with boron-epoxy composite elements replacing the aluminum skins. In both cases the structure is designed with and without displacement constraints. In the displacement constrained design, the maximum deflection at a node point is restricted to 36 inches. The results of these problems are given in Table 13. For the design without displacement constraints, the number of laminae in each fiber direction is given in Figure 16. The 0^0 fibers are parallel to the span of the wing. For the design without displacement constraints, the maximum deflection was found to be 80.02 inches and 43.28 inches at node point 1 in the Z direction, for the aluminum and composite wing respectively. It is seen from Table 13 that the composite wing is 46% lighter than the aluminum wing.

Design Examples with Static Stability Requirements

The remainder of the problems are concerned with static stability requirements. The selected structures are such that the sizes of the members are governed by the buckling load on the structure and not by stresses in individual members. The columns, transmission towers, and tall building frames are some of the structures which fall under this category. The recurrence relation used in resizing members in each iterative cycle
is given in section 3. For additional details the reader is referred to

[10].

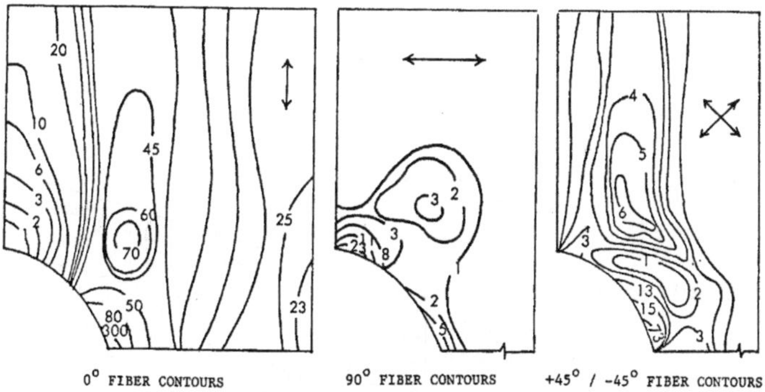

FIGURE 14. FIBER ORIENTATION CONTOURS FOR A PLATE WITH A HOLE

Table 13. Weights and deflections of the cantilever wing

	ISOTROPIC			COMPOSITE		
	Initial Weight	Final Weight	Max Defl	Initial Weight	Final Weight	Max Defl
Stress Limited	5982.6	4905.5	80.02	6102.1	2596.0	43.28
Displ Limited	16299.9	10410.0	36.00	12046.3	3091.1	36.00

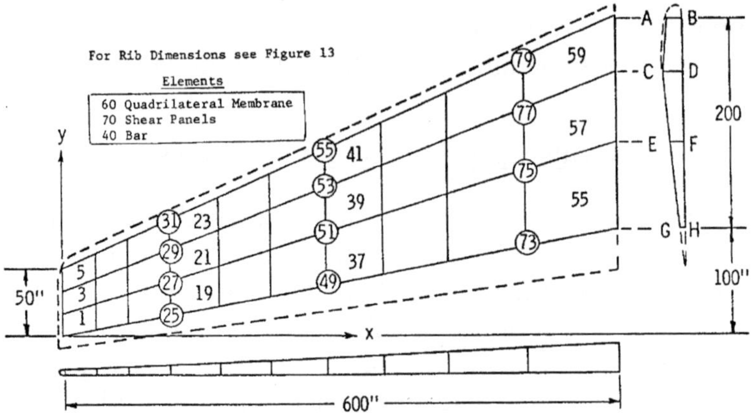

Figure 15. Finite Element Model of the Cantilever Wing

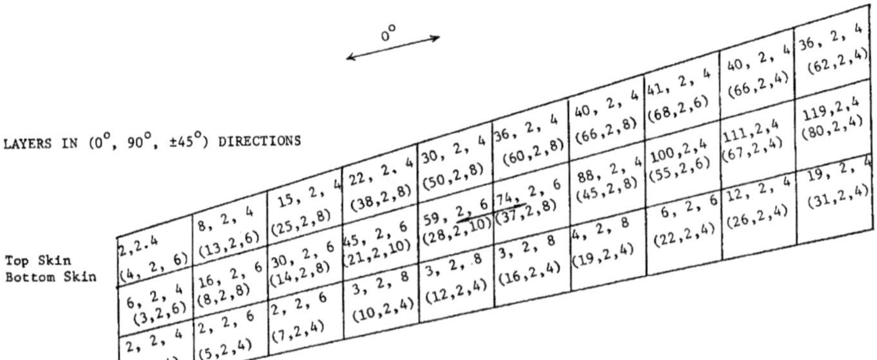

Figure 16. Optimum Design of the Cantilever Wing

Table 14. Weights of Columns

	free fixed	hinged hinged	hinged fixed	fixed fixed
Initial Weight	351.74	87.93	343.81	175.82
Final Weight	291.45	74.14	292.50	150.21
INITIAL / FINAL	1.2068	1.1859	1.1754	1.1704

Table 15. Relative Areas of Minimum Weight Columns

Element No.	free fixed	hinged hinged	hinged fixed	fixed fixed
1	0.0945	0.1826	0.2632	1.0000
2	0.2706	0.4669	0.6594	0.6360
3	0.4310	0.7188	0.9101	0.1916
4	0.5784	0.9030	1.0000	0.6361
5	0.7003	1.0000	0.9286	0.9998
6	0.8010	1.0000	0.6963	0.9998
7	0.8811	0.9030	0.3126	0.6361
8	0.9407	0.7188	0.2058	0.1916
9	0.9803	0.4669	0.5920	0.6360
10	1.0000	0.1826	0.8464	1.0000

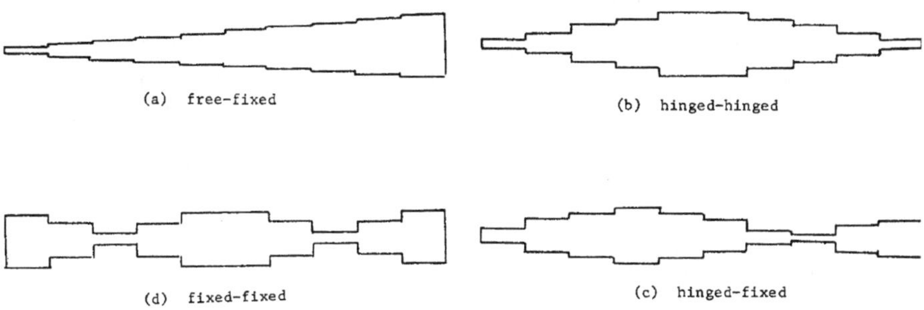

FIGURE 17. Material Distribution in Optimized Columns

Example 12: Columns with Different Boundary Conditions

The column is designed for four boundary conditions. The boundary conditions are free-fixed, hinged-hinged, hinged-fixed and fixed-fixed. The column is divided into ten finite elements and is subjected to an axial load of 10 kips, and the cross-section of the column is assumed to be hollow and rectangular(2"x4"). For convenience, the ratio of area to moment of inertia is assumed to be constant so that the depth (4") of all elements is constant. This allows scaling the design to carry a specified buckling load without additional analysis cycles. The column is not allowed to buckle out of the plane of bending. The length of the column for first two boundary conditions is 25' and 50' for the remaining two boundary conditions.

For the last two boundary conditions, this increase is desirable to make the column critical in buckling. The initial and final weights of the columns are given in Table 14, The initial weight is calculated on the basis of uniform inertia and area for all elements. The final designs given in Table 14 are arrived at after a couple of iterations. The ratio of initial to final weight is also given in Table 14. For the free-fixed and hinged-fixed cases this ratio for the continuously varying section is found to be 1.216 and 1.225. The difference between these values and those given in Table 14 appear to be due to the finite element idealization. Table 15 gives the relative areas of different elements of the columns, and the same values are plotted in Figure 17 to show the variation for different boundary conditions. Since it is assumed that the ratio of cross-sectional area to moment of inertia is constant for all the elements, the distribution of area as well as inertia will be the same as shown in Figure 17. It is seen from this figure that for the hinged-hinged and fixed-fixed cases, the distribution is symmetric about the midpoint of the column. This is expected because of symmetric boundary conditions. Additional problems may be found in [10].

VI. Summary and Conclusions

This paper has attempted in Section 2 to present a unified approach to the derivation of optimality criteria for a variety of structural optimization problems of practical interest. For each of the design requirements an iterative algorithm is derived in Section 3. Some discussion of the problems of multiple minima and the approximate nature of the design methods is presented with the aid of a two variable design space in Section 4. A number of design examples with static, dynamic and stability requirements are presented in Section 5. These examples include isotropic and layered fiber reinforced composite structures.

The iterative algorithms derived in Section 3 assure reasonably stable convergence to the optimum. The number of design cycles is usually not dependent on the number of design variables in the problem. The rapid convergence of the optimality criteria approach makes it economically feasible to automate the design procedure.

VII. References

1. Gellatly, R., "Development of Procedures for Large Scale Automated Minimum Weight Structural Design", AFFDL-TR-66-180.

2. Venkayya, V.B., Khot, N.S., and Reddy. V.S., "Energy Distribution in an Optimum Structural Design", AFFDL-TR-68-156.

3. Venkayya, V.B., "Design of Optimum Structures", Journal of Computers and Structures, Vol. 1, p. 265309, Pergamon Press, August 1971. Also presented in conference on Computer Oriented Analysis of Shell Structures, Palo Alto, California, August 1970, AFFDL-TR-71-79.

4. Berke, L., "An Efficient Approach to the Mininum Weight Design of Deflection Limited Structures". AF Flight Dynamics Laboratory Technical Memorandum, TM-70-4-FDTR (July 1970).

5. Dwyer, W.J., Emerton, R.K., and Ojalvo, I.U., "An Automated Procedure for the Optimization of Practical Aerospace Structures" (Volume I), AFFDL-TR-70-118.

6. Prager, W., and Marcal, P.V., "Optimality Criteria in Structural Design", AFFDL-TR-70-166.

7. Gellatly, R.A., and Berke L., "Optimum Structural Design," AFFDL-TR-70-165.

8. Venkayya, V.B., Khot, N.S., Tischler, V.A., and Taylor, R.F., "Design of Optimum Structures for Dynamic Loads," 3rd Air Force Conference on Matrix Methods in Structural Mechanics, Oct. 1971.

9. Khot, N.S., Venkayya, V.B., Johnson, C.D., and Tischler, V.A., "Application of Optimality Criterion to Fiber Reinforced Composites," AFFDL-TR-73-6.

10. Khot, N.S. and Venkayya, V.B., "Optimization of Structures for Strength and Stability Requirements," AFFDL-TR-73- in preparation.

11. Berke, L., "Convergence Behavior of Iterative Resizing Procedures Based on Optimality Criteria" AFFDL-TM-72-1-FBR.

12. Taig, I.C. and Kerr, R.I., "Optimisation in Aircraft Structures", Symposium on Optimisation in Aircraft Design, Royal Aeronautical Society, November 1972.

13. Schmit, L.A., Structural Design by Systematic Synthesis. 2nd Conf. Electronic Computation, ASCE Structural Division, Pittsburgh, Pennsylvania, pp. 105-132 (1960).

References

Al-Banna, S. (1972), "Iterative Structural Design", Ph.D. thesis, Columbia University, New York City.

Allaire, G., Jouve, F., and Toader A-M., (2004), "Structural Optimization Using Sensitivity Analysis and a Level-set Method", *Journal of Computational Physics,* 194, pp. 363–393.

Arbel, A. (1993), *Exploring Interior-Point Linear Programming,* MIT Press, Cambridge, MA.

Balling, R. J., Briggs, R. R. and Gillman K. (2006), "Multiple Optimum Size/Shape/Topology Designs for Skeletal Structures Using a Genetic Algorithm", *Journal of Structural Engineering,* 132, 7, pp. 1158–1165.

Bathe, K. J., and Wilson, E. L. (1973), "Eigensolution *of* Large Structural Systems with Small Bandwidths," *Journal of the Structural Division,* ASCE, Vol. 99. No. ST6, pp. 467–479.

Beckenbach, E. F., and Bellman, *R..* (1971), *Inequalities.* Springer-Verlag, New York.

Bendsoe, M. P., and Mpta Soares C. A. (1992), *Topology Design of Structures,* Kluwer Academic Publishers, The Netherlands.

Bixby, R. E., Gregory, J. W., Lustig, I. J., Marsten, R. E., and D. F. Shanno (1991), "Very Large Scale Linear Programming: A Case Study in Combining Interior Point and Simplex Methods", Rutcor Research Report #34-91, Rutgers University.

Brandt, S. C., Morbach, J., Miatidis, M., Theissen, M., Jarke, M., and Marquardt W., (2008), "An Ontology-Based Approach to Knowledge Management in Design Processes", *Computers and Chemical Engineering,* 32,1–2, pp. 320–342.

Byrd, R. H., Gould, N. I. M., Nocedal, J., and Waltz R. A. (2005), "On the Convergence of Successive Linear-Quadratic Programming Algorithms", *SIAM Journal of Optimization,* 16, 2, pp.471–488.

Camp, C. V., and B. J. Bichon (2004), "Design of Space Trusses Using Ant Colony Optimization", *Journal of Structural Engineering,* 130, 5, pp. 741–751.

Cassis. J. H., and Schmit, L. A.. Jr. (1976), "Optimum Structural Design with Dynamic Constraints," *Journal* of *the Structural Division.* ASCE. 102. No. STIO, pp. 2053–2071.

Chan, C-M. (1997), "How to Optimize Tall Steel Building Frameworks", *Guide to Structural Optimization,* ASCE Manuals and Reports on Engineering Practice No. 90, pp. 165–196.

Chan, C-M. and J. K. L. Chui (2006) "Wind-Induced Response and Serviceability Design Optimization of Tall Steel Buildings", *Engineering Structures,* 28, pp. 503–513.

Chern, N. J., Prager, W. (1971), "Minimim Weight Design of Statically Determinate Trusses Subject to Multiple Constraints", *International Journal of Solids and Structures*, 7.

Choi, C-K., Lee, D-G., and Lee H-W., (1991), "Optimization for Large-Scale Steel Structures with Discrete Sections", *Computers and Structures,* 39, 5, pp. 547–556.

Connor, J. J. (2003), *Introduction to Structural Motion Control,* Prentice Hall, Upper Saddle River, NJ.

Courant, R., and D. Hilbert (1953), *Mathematical Methods of Physics, I,* Interscience Publishers, New York City.

Damkilde, L., Hoyer, O., and Krenk, S. (1994), "A Direct Linear Programming Solver in C for Structural Applications", *Computers and Structures,* 52, 3, pp. 511–528.

Den Hartog, J. P. (1985), *Mechanical Vibrations,* Dover Publications, New York City.

Denoon, R. (2006), "Dampers – Do We Really Need Then?", *Structure Magazine,* June 2006, pp. 24–26.

Du, H., and Qin H., (2004), "A Shape Design System Using Volumetric Implicit PDSs", *Computer-Aided Design,* 36, pp.1101–1116.

Duffin, R. J., Peterson, E. L., and Zener C. (1967), *Geometric Programming: Theory and Application,* John Wiley and Sons, New York.

Ecker, J. G., and Kupferschmid M. (1991), *Introduction to Operations Research,* Krieger Publishing Company, Malabar, Florida.

Ehrgott, M. (2005), *Multicriteria Optimization,* Springer Verlag, Berlin.

El Hallabi, M. (1994), "A Global Convergence Theory for Sequential Linear Programming Inexact Hybrid Algorithms", CRPC-TR94371, Rice University Center for Research on Parallel Computation.

Farkas, J. (1984), *Optimum Design of Metal Structures,* Akademiai, Budapest.

Farkas, J., and Jarmai K. (1997), *Analysis and Optimum Design of Metal Structures,* A. A. Balkema, Rotterdam.

Fitfler, E. (2001), "Evolutionary Algorithms for Multiobjective Optimization", Eurogen 2001 (invited lecture).

Fox, R. L. (1969), "Mathematical Methods in Optimization" in *An Introduction to Structural Optimization,* M. Z. Cohn (editor), Study #1, Solid Mechanics Division, University of Waterloo, Ontario.

Fox, R. L., and Kapoor. M. P. (1968), "Rates of Change of Eigenvalues and Eigenvectors" *AIAA Journal,* 6, 12. Dec., pp. 2426–2429.

Funaro, J. (1974), "Optimal Structural Design with Deflection Constraints", Ph.D. thesis, Columbia University, New York City.

Fylstra, D., Lasdon, L., Watson, J., and Waren, A. (1998), "Design and Use of the Microsoft Excel Solver", *Interfaces*, 28, 5, pp. 29–55.

Gale, D. (1960), *The Theory of Linear Economic Models,* McGraw-Hill, New York.

Gero, J. S. (editor) (1985), *Design Optimization,* Academic Press, New York.

Gero, J. S., and M. L. Maher (2005) *Computational and Cognitive Models of Creative Design VI,* Key Centre of Design Computing and Cognition, University of Sydney, Australia. (There is a remarkable volume of material relating to architectural design that can be found on Professor Gero's website (http://mason.gmu.edu/~gero//publications/2008.html/))

Haftka, R. T., Gurdal, Z., and Kamat, M. P. (1990), *Elements of Structural Optimization,* Kluwer Academic Publishers, Dordrecht.

Hagishita, T., and Ohsaki M. (2008), "Topology Optimization of Trusses by Growing Ground Structure Method", *Structural and Multidisciplinary Optimization,* DOI10.1007/s00158-008-0237-4.

Harrison, M. A. (1965), *Introduction to Switching and Automata Theory,* McGraw-Hill, New York, NY.

Hartfield, R. J., Jenkins, R. M., and Burkhalter J. E. (2007), "Ramjet Powered Missile Design Using a Genetic Algorithm", *Journal of Computing and Information Science in Engineering,* 7, pp. 167–173.

Hemp, W. S. (1968) Abstract of Lecture Course "Optimum Structures", 2nd Ed., Engineering Laboratory, University of Oxford. (This reference is due to G. G. Pope, "Classical Optimization Theory Relevant to the Design of Aerospace Structures" in *Structural Design Applications of Mathematical Programming Techniques*, G. G. Pope and L. A. Schmidt, editors, AGARDograph #149, NASA, Langley Field, Virginia, 1972.)

Heyman, J. (2008), *Basic Structural Theory,* Cambridge University Press, Cambridge.

Hodge, P. (1954), *Plastic Analysis of Structures*, McGraw Hill, New York.

Isaacson, E., and Keller H. B. (1966), *Analysis of Numerical Methods,* John Wiley & Sons, New York City.

Jen, E. (editor) (2005), *Robust Design,* Cambridge University Press, New York, NY.

Kang, B-S., Park, G-J., and Arora J. (2006), "A Review of Optimization of Structures Subjected to Transient Loads", *Structural and Multidisciplinary Optimization,* 31, pp. 81–95.

Kaplan, W. (1952), *Advanced Calculus,* Addison-Wesley Company, Cambridge, MA.

Kaplan, W. (1953), *Advanced Calculus,* Addison-Wesley, Cambridge.

Karmarkar, N. K. (1984), "A New Polynomial Time Algorithm for Linear Programming", *Combinatorica,* 4, pp. 373–395.

Keller, J. B. (1960), "The Shape of the Strongest Column", *Archive for Rational Analysis and Mechanics,* 5, pp. 275–285.

Khan, M. R. (1984), "Optimality Criteria Techniques Applied to Frames Having General Cross-Sectional Relationships", *AIAA Journal,* 22, 54, pp. 669–676.

Kirsch, U. (1981), *Optimum Structural Design,* McGraw-Hill, New York City.

Knight, T. W. (1994), *Transformations in Design,* Cambridge University Press, Cambridge, UK.

Koumousis, V. K., and Georgiou, P. G. (1994), "Genetic Algorithms in Discrete Optimization of Steel Truss Roofs", *Journal of Computing in Civil Engineering,* 8, 3, pp. 309–325.

Kumar, A. V. (2000), "A Sequential Optimization Algorithm Using Logarithmic Barriers: Applications to Structural Optimization", *Journal of Mechanical Design,* 122, 3, pp. 271–277.

Lamberti, L., and Pappalettere, C. (2000), "Comparison of the Numerical Efficiency of Different Sequential Linear Programming Based Algorithms for Structural Optimization Problems", *Computers and Structures,* 76, pp. 713–728.

Lamberti, L., and Pappalettere, C. (2004), "Improved Sequential Linear Programming Formulation for Structural Weight Minimization", *Computer Methods in Applied Mechanics and Engineering,* 193, pp. 3493–3521.

Lee, C-L., Chen, Y-T., Chung, L-L., and Wang Y-P., (2006), "Optimal Design Theories and Applications of Tuned Mass Dampers", *Engineering Structures,* 28, pp. 43–53.

Lev, O., and Spillers W. R. (1971), "Design for Two Loading Conditions", *International Journal of Solids and Structures,* 7, pp. 1261–1267.

Levy, R., and Spillers W. R. (2003), *Analysis of Geometrically Nonlinear Structures,* Kluwer Academic Publishers, Dordrecht.

Lorch, E. R. (1951), "Differential Inequalities and the Theory of Convex Bodies", Transactions of the American Mathematical Society, 71, 2, pp. 243–266.

Luenberger, D. G. (1984), *Linear and Nonlinear Programming,* Addison Wesley, Reading, MA.

McCormac, J. C. (2007) *Structural Steel Design,* Prentice Hall, Upper Saddle River, NJ.

Michell, A. G. M. (1904), "The Limits of Economy in Frame-structures", *Philosophical Magazine, S. 6.,* 8, 47.

Michalski, R. S., and Kaufman K. A. (2006), "Intelligent Evolutionary Design: A New Approach to Optimizing Complex Engineering Systems and Its Application to Designing Heat Exchangers", *International Journal of Intelligent Systems,* 21, pp. 1217–1248.

Morris, A. J. (1972), "Structural Optimization by Geometric Programming", *International Journal of Solids and Structures,* 8, pp. 847–864.

Newland, D. E. (1986), *An Introduction to Random Vibrations and Spectral Analysis,* Longman, Harlow, England.

Oberlin, C., and Wright S. J., (2006), "Active Set Identification in Nonlinear Programming", *SIAM Journal of Optimization,* 17, 2, pp. 577–605.

Onat, E. T., Schumann, W. and Shield R. T. (1957), "Design of Circular Plates for Minimum Weight", *Journal of Applied Mathematics and Physics,* VIII, 6, pp. 485–499.

Pennington, S. V., and Berzins M. (1994), "New NAG Library Software for Partial Differential Equations", *ACM Transactions on Mathematical Software*, 20, 1, pp. 63–99.

Pfaltz, J. L., and Rosenfeld, A. (1969), "Web Grammars", *Proceedings of the International Joint Conference on Artificial Intelligence*, Washington D.C., pp. 609–619.

Queipo, N. V., Haftka, R. T., Shyy, W., Goel, T., Vaidyanathan, R., and Tucker P. K. (2005), "Surrogate-Based Analysis and Optimization", *Progress in Aerospace Sciences,* 41, pp. 1–28.

Rozvany, G. I. N. (1976), *Optimal Design of Flexural Systems,* Pergamon Press, Oxford.

Rozvany, G. I. N., and Adidam, S. R. (1972), "Absolute Minimum Volume of Reinforcement of Slabs", *Journal of the Structural Division*, ASCE, 98, ST5, pp. 1217–1222.

Rozvany, G. I. N., Bendsoe, M. P., and Kirsch U. (1995), "Layout Optimization of Structures", *Applied Mechanics Reviews,* 48, pp. 41–119. (This reference serves as a good pointer to Rozvany's work.)

Sahinidis, N. V. (2004), "Optimization Under Uncertainty: State-of-the-Art and Opportunities", *Computers and Chemical Engineering,* 28, pp. 971–983.

Sarma, K. C., and Adeli H. (2000), "Cost Optimization of Steel Structures", *Engineering Optimization,* 32, 6, pp. 777–802.

Save, M. and Prager W. (1985), *Structural Optimization, Volume 1, Optimality Criteria,* Plenum Press, New York.

Schiesser, W. E. (1991), *Numerical Method of Lines,* Academic Press, San Diego, CA.

Schittkowski, K., Zillober, C., and Zotemantel R. (1994), "Numerical Comparison of Nonlinear Programming Algorithms for Structural Optimization", *Structural Optimization, 7,* pp.1–19.

Schmit, L. A. (1960), "Structural Design By Systematic Synthesis", *Proceedings of the Second Conference on Electronic Computation,* ASCE, pp. 105–122.

Shield, R. T. (1959), "On the Optimum Design of Shells", *Journal of Applied Mechanics,* paper #59 A-47, pp. 1–7.

Shneiderman, B. (2007), "Creativity Support Tools: Accelerating Discovery and Innovation", *Communications of the ACM,* 50, 12, pp. 20–32.

Simonnard, M. (1966), *Linear Programming,* Prentice-Hall, Englewood Cliffs, NJ.

Sivakumar, P., Rajaraman, A., Samuel Knight, G. M., and Ramachandramurthy D. S. (2004), "Object-Oriented Optimization Approach Using Genetic Algorithms for Lattice Towers", *Journal of Computing in Civil Engineering,* 18, 2, pp. 162–171.

Soondal, P. The linear programming solver LPAKO is described completely at http://www.ici.ro/camo/linear/lpako.htm.

Spillers, W. R. (1972), *Automated Structural Analysis: An Introduction,* Pergamon Press, New York. NY. (1972).

Spillers, W. R. (1975) *Iterative Structural Design,* North Holland, Amsterdam. (This book is available online at http://web.njit.edu/civil/fabric/optim/indx2.html)

Spillers, W. R. (1985), *Introduction to Structures,* Ellis Horwood, Chichester.

Spillers, W. R., and Farrell. J. (1969), "An Absolute Value Linear Programming Problem." *Journal of Mathematical Analysis and Applications,* 28, 1. Oct, 1969 pp. 153–158.

Spillers, W. R., and Kountouris G. E., (1980), "Geometric Optimization Using Simple Code Representations", *Journal of the Structural Division, Proceedings of the American Society of Civil Engineers,* 106, ST5, pp. 959–971.

Spillers, W. R., Singh, S., and Levy, R. (1981a), "Optimization with Frequency Constraints", *Journal of the Structural Division, ASCE,* 107, ST12, pp. 2337–2347.

Spillers, W. R., Singh, S., and Levy R. (1981b), "Optimization with Frequency Constraint", *Proceedings ASCE,* 107, ST12, pp. 2337–2346.

Stiny, G. (2006), *Shape,* MIT Press, Cambridge, MA.

Symonds, P. S., and Prager, W. (1950), "Elastic-Plastic Analysis of Structures Subjected to Load Varying Arbitrarily Between Prescribed Limits", *Journal of Applied Mechanics*, pp. 315–323.

Tabak, E., and Wright P. M. (1981), "Optimality Criteria Method for Building Frames", *Journal of the Structural Division, ASCE,* 107, ST7, pp. 1327–1342.

Timoshenko, S. (1936), *Theory of Elastic Stability*, McGraw-Hill, New York.

Umesha, P. K., Venuraju, M. T., Hartmann, D., and Leimbach K. R. (2007), "Parallel Computing Techniques for Sensitivity Analysis in Optimum Structural Design", *Journal of Computing in Civil Engineering,* 21, 6, pp. 463–477.

van de Lindt, J. W., and Dao T. N. (2007), "Evolutionary Algorithm for Performance-Based Shear Wall Placement in Buildings Subjected to Multiple Load Types", *Journal of Structural Engineering,* 133, 8, pp. 1156–1167.

Vanderplaats, G. N., and Moses F. (1972), "Automated Design of Trusses for Optimum Geometry", *Journal of the Structural Division, Proceedings of the ASCE,* 98, ST3, March 1972.

Venkayya, V. B. (1978). "Survey of Optimization Techniques in Structural Design," *Technical Memorandum TM-FBR-28-43,* Wright-Patterson AFB, Ohio.

Venkayya V. B. (1971), "Design of Optimum Structures", *Computers and Structures,* 1, pp. 265–309.

Venkayya, V. B., Khot, N. S., and Berke L. (1973), "Application of Optimality Criteria Approaches to Automated Design of Large Practical Structures", Second Symposium on Structural Optimization, AGARD-CP-123, Milan, Italy.

Visser, W. (2006), "Designing as Construction of Representations: A Dynamic Viewpoint in Cognitive Design Research", *Human-Computer Interaction,* 21, pp. 103–152.

Wang, D., Zhang, W. H., and Jiang, J. S. (2002), "Truss Shape Optimization with Displacement Constraints", *Computer Methods in Applied Mechanics and Engineering,* 191, pp. 3597–3612.

Wasiutynshi, Z. and Brandt A. (1963), "The Present State of Knowledge in the Field of Optimum Design of Structures", *Applied Mechanics Reviews,* 16, 5, pp. 341–350.

Wouwer, A. V., Sausez, Ph., and Schiesser W. E. (2001), *Adaptive Method of lines,* Chapman & Hall/CRC, Boca Raton, FL.

Wright, S. J. (1997), *Primal-Dual Interior-Point Methods,* SIAM Press, Philadelphia.

Young, J. W., Jr., and Christiansen. H. M. (1966), "Synthesis of a Space Truss Based on Dynamic Criteria," *Journal* of *the Structural Division,* ASCE. Vol. 92. No. ST6. Dec., 1966. pp. 425–442.

Zhou, Y., Kijewski, T., and Kareem A. (2003), "Aerodynamic Loads on Tall Buildings: Interactive Database", *Journal of Structural Engineering,* 129, 3, pp. 394–404.

Index

Springer

MIX
Papier aus verantwortungsvollen Quellen
Paper from responsible sources
FSC® C105338

If you have any concerns about our products,
you can contact us on
ProductSafety@springernature.com

In case Publisher is established outside the EU,
the EU authorized representative is:
Springer Nature Customer Service Center GmbH
Europaplatz 3, 69115 Heidelberg, Germany

Printed by Libri Plureos GmbH
in Hamburg, Germany